TENDERS AND CONTRACTS
FOR BUILDING

Other titles by the Aqua Group

Pre-Contract Practice for Architects and Quantity Surveyors

Contract Administration for the Building Team

Fire and Building

TENDERS AND CONTRACTS FOR BUILDING

SECOND EDITION

THE AQUA GROUP

With sketches by
Brian Bagnall

OXFORD

BLACKWELL SCIENTIFIC PUBLICATIONS

LONDON EDINBURGH BOSTON

MELBOURNE PARIS BERLIN VIENNA

Copyright © The Aqua Group 1975,
    1982, 1990

Blackwell Scientific Publications
Editorial Offices:
Osney Mead, Oxford OX2 0EL
25 John Street, London WC1N 2BL
23 Ainslie Place, Edinburgh EH3 6AJ
238 Main Street, Cambridge
    Massachusetts 02142, USA
54 University Street, Carlton
    Victoria 3053, Australia

First published in 1975 by Crosby
    Lockwood Staples (under the title 'Which
    Builder?'; original ISBN 0 258 97041 3)
Published in 1982 by Granada Publishing
    (under the title 'Tenders and Contracts
    for Building'; ISBN 0 246 118385)
Reprinted 1984
Reprinted by Collins Professional and
    Technical Books 1986
Second edition published by BSP
    Professional Books 1990
Reprinted by Blackwell Scientific
    Publications 1992, 1994

Set by Kudos Graphics, Horsham,
West Sussex
Printed and bound in Great Britain by
Hartnolls Ltd, Bodmin, Cornwall

DISTRIBUTORS

Marston Book Services Ltd
PO Box 87
Oxford OX2 0DT
(*Orders:* Tel: 0865 791155
        Fax: 0865 791927
        Telex: 837515)

USA
Blackwell Scientific Publications, Inc.
238 Main Street,
Cambridge, MA 02142
(*Orders:* Tel: 800 759-6102
            617 876-7000)

Canada
Oxford University Press
70 Wynford Drive
Don Mills
Ontario M3C 1J9
(*Orders:* Tel: (416) 441-2941)

Australia
Blackwell Scientific Publications Pty Ltd
54 University Street
Carlton, Victoria 3053
(*Orders:* Tel: 03 347-5552)

British Library
Cataloguing in Publication Data

Tenders and contracts for building.
2nd ed.
1. Great Britain. Buildings.
    Construction. Tendering
2. Great Britain. Buildings. Construction.
    Contractual arrangements
I. Aqua Group
692.50941

ISBN 0-632-02681-2

Contents

INTRODUCTION

Such is the rate of change in building procedures and contracts that our trilogy of books on the subject continues under almost constant review. As this, the last of the series, comes into its second edition, *Pre-Contract Practice* will shortly go into its seventh edition and *Contract Administration* has just appeared in its seventh edition. This book's original title, *Which Builder? Tendering and Contractual Arrangements*, was dropped in favour of *Tenders and Contracts for Building*, a simpler explanation of a subject that is in fact more complicated than might be supposed.

Pre-Contract Practice and *Contract Administration* are concerned with good practice in preparing for and administering building contracts. In both books emphasis is placed on the fact that the recommended procedures are based on work carried out under the JCT Standard Form of Building Contract following the selection of a contractor by competitive tender. There are many occasions, however, when it is advantageous to appoint the contractor not as a consequence of a normal competitive tendering procedure, but by following a different set of procedures more suited to the scope of work, its programme or existence of other contracts.

It is the various tendering procedures, contractual arrangements and the circumstances under which they might be adopted that are examined in this book. In particular the main principles are set out for consideration so that the most advantageous choice may be made.

In a number of reports of recent years the National Economic Development Office has also put forward a wide range of alternative contractual procedures. Many of their recommendations coincide with those made in this book. This applies particularly to the emphasis on continuity, the use of a wide range of procedures in tendering for public work and a need for a clearer and more balanced view of public accountability. In this edition Chapter 3 on Accountability has been elaborated upon to present such a view.

Since publication of the first edition of this book the JCT Form of Management Contract has been introduced and a new chapter has been introduced to describe its use (Chapter 11).

The order of chapters has in fact been rearranged into a more logical sequence. Chapter 4, Early Selection: Two-Stage Tendering, was originally Chapter 8 and has been rewritten to provide a more thorough analysis and introduction to that subject. Chapter 5, Drawings, Specification and Quantities (originally Chapter 6) now concentrates on the make-up and status of tender documents, whilst a new Chapter 8 on Fixed Price Contracts has been introduced.

The remaining chapters have all been extended and updated. Greater attention is now given to the appraisal of procedures, their advantages, disadvantages and suitability to the occasion. Appendix B listing the Standard Forms of Building Contracts has been updated. Explanatory diagrams have been introduced.

On a point of detail, for the sake of clarity it has been decided to use the term 'client' throughout the book rather than 'employer' which is the term used in all the forms of contract.

Our appreciation is due to the past chairman of the Aqua Group, Tony Brett-Jones, who was primarily responsible for the original book and whose regard for the spirit of the contract has provided guidance to those responsible for this revised edition. Once again Brian Bagnall, with his now classic illustrations, has introduced that touch of humour that so helps things along. We thank him for it.

The Aqua Group comprises:

Brian Bagnall BArch (Liverpool)
Tony Brett-Jones CBE, FRICS, FCIArb
Helen Dallas Dip Arch, RIBA
Peter Johnson FRICS, FCIArb
John Oakes FRICS, FCIArb
Richard Oakes FRICS
Quentin Pickard BA, RIBA (Chairman)
Geoffrey Poole FRIBA, ACIArb
Geoff Quaife ARICS
Colin Rice FRICS
John Townsend FRICS, ACIArb
John Willcock Dip Arch, RIBA
James Williams Dip Arch (Edin), FRIBA

'. . . *proper time* . . .'

Chapter 1

Making the Contract

Like any other commercial contract a building contract is made on the basis of an offer and an acceptance for a consideration to be paid from one to the other party.

This book is concerned with that operation, the types of contract conditions which are appropriate in any given circumstances and the various ways in which the consideration may be arrived at.

The majority of building contracts are relatively small and are probably conducted on quite an informal basis. A householder will call in the local builder about some alterations he wants done, explain his requirements to the builder, perhaps obtain an estimate, and then tell the builder to get on with the work. The estimate and the acceptance may be verbal or in writing, or there may be no estimate at all – just an implied agreement that the builder will be paid for what he does. Such contracts – for contracts they are – are everyday matters in the domestic and minor works field. It is not our intention to deal with them further in this book.

It is proposed to deal with the larger works in which formal procedures and proper documentation play an important part, perhaps because of the amount of money involved, perhaps because of the nature or complexity of the work, and perhaps because of the needs of financial control, accountability or the public interest.

The process of making the contract, of whatever size, may be divided into three parts:

(1) the decision on the type of contract and the particular contract conditions and documentation under which the work will be carried out
(2) the selection of the contractor
(3) the establishment of the contract price or how the price will be arrived at.

Contract conditions

The choice of the conditions or the form of contract to be used will normally be made by the client in the light of advice he receives from his professional advisers. A choice must be made early on as it will affect the way in which the supporting documents are prepared. The type of contract and the actual conditions to be used must also be defined before the contractor can be selected and before the contract price or method of calculating it can be settled.

The options available are numerous and the types of contract in common use may be summarised as follows:

(1) fixed price contracts, with or without fluctuations, based on:
performance specification
specification and drawings
schedule of rates
bills of quantities
bills of approximate quantities
(2) cost reimbursement contracts
prime cost + percentage
prime cost + fixed fee
(3) target cost contracts
(4) management contracts
(5) design/build contracts
(6) continuity contracts.

Tendering procedure

The selection of the contractor and the establishment of the contract price, or how the price will be arrived at, may be combined in a single operation or they may be two separate operations.

First it is helpful to look at a definition of tendering that has held good for a number of years:

> *The purpose of any tendering procedure is to select a suitable contractor, at a time appropriate to the circumstances, and to obtain from him at the proper time an acceptable tender or offer upon which a contract can be let.*

In fact, this is not so much a definition as a statement of the purpose of tendering. However, it needs amplification for a full understanding of all that it covers. The statement emphasises that the purpose of tendering is the selection of the contractor and the obtaining of the tender or offer.

Although any tender may take account of the conditions of contract under which the work will be performed, the tendering procedure is not dependent on the type of contract to be used. The contract conditions have the same significance as, for example, contract drawings and bills of quantities, which together make up the total contractual arrangements. It is important to appreciate this, otherwise considerable confusion will prevail. Contractual arrangements are concerned with the type of contract to be entered into and the obligations, rights and liabilities of the parties to the contract. These may vary because of the type of project, but there is no direct relationship between them and the tendering procedure.

For example, the same contractual arrangements could prevail in two cases where, in one, the contractor had been selected after submission of a competitive tender based on detailed drawings and the bills of quantities, while in the other, he had been selected on the basis of a business relationship with the client and the single tender negotiations.

Having identified the differences between tendering procedures and contractual arrangements, let us consider in more detail the definition of tendering procedures. One of the first things to note is that the word time occurs twice. On the first occasion it refers to the 'time appropriate' to select a suitable contractor. The second reference is to the 'proper time' to obtain a tender. In other words, the time for selection of the contractor may well be different from the time at which one obtains a tender or establishes the contract price. A further point to note is that the acceptable tender or offer does not finish the matter, but is the basis upon which a contract can be let.

It may be thought that we have examined an apparently simple definition in rather too much detail. This is deliberate and the succeeding chapters will give consideration to the various facets of this definition.

There has in the past been an underlying assumption that the tendering process was only concerned with finding out which contractor could submit the lowest price based on the design presented to him. Only in recent years has greater consideration been given to situations where the contractor is partly or wholly involved in the design; where circumstances require that construction commences before the design has been completed; or where, for any other reason, price and time alone may not be the only factors to be taken into account.

In considering the purpose of tendering, one must think in the broadest terms. A tender quotes not just a price but also a time in which to complete the work. The purpose of tendering may not be for one job; it may be necessary to consider it in terms of a total programme, of which that job is just one project.

Tendering procedure today is a dynamic situation. It is not a procedure which can be applied across the board on all construction works; in fact, 'procedure' is probably the wrong word, but as it is used so frequently it may be misleading not to use it here. There are so many different factors to consider in each case, and hence decisions to be made, that it would be more appropriate to refer to it as the tendering structure and requirement.

Much of what has been written with regard to past attitudes to tendering procedures also applies to contractual arrangements. There has been great emphasis on fixed price contracts with a somewhat poor realisation as to what the word fixed entailed. Frequently the assumption has been that the contract should be a fixed price one, unless the situation is such as to make this impossible.

Although the actual form and the conditions of contract are part of the contractual arrangements, we are not concerned in this book with the legal contract as such. Numerous standard forms exist and a list of those in common use is given in Appendix B. The choice of the actual form to be used on any given project is a matter to be decided in relation to the nature of the work and the needs of the client, and is not a matter which will be considered in this book. Our concern relates to the general contractual arrangements, whether on a fixed price, cost reimbursement or target cost basis.

Finally, although everything written here is in terms of a contract between the client and the main contractor, it is equally applicable to a contract between a main contractor and a sub-contractor.

'. . . the risk inherent in the construction . . .'

Chapter 2

General Principles

Chapter 1 dealt with the aim and purpose of tendering and contractual arrangements. This chapter deals with some of the general principles that have to be considered in selecting a contractor and formulating arrangements whereby a satisfactory contract may be made.

It has been said that time and money are the only considerations in selecting a contractor and making a contract with him. While this is true in broadest terms, there are many factors which affect either time or the financial outcome of the contract but which are not, at first, apparent as being directly related to either.

Economic use of resources

It may be thought that the money that is paid to the contractor is all that has to be considered. Money represents the resources that the contractor has to use. Thus a contractor may quote the lowest price of those who tender and yet the client can still have an unsatisfactory situation. If the tendering procedure and contractual arrangements are badly drawn up, the contractor may be using his resources inefficiently. This is dealt with in much greater detail later. To give a very simple example here: if a client has two similar projects, he will not necessarily get the lowest final price by setting up competition between one set of contractors for the first project and another set for the second and taking the lowest in each case. If the two projects are combined in one, the price of the larger single contract may be less than the two let separately, because of the savings in building resources.

It is our view that the first priority in tendering is to ensure the most economic use of the building resources, bearing in mind the particular needs of the client, and then to ensure that the price paid for those resources is as low as is reasonably possible.

The economical use of contractors' resources is becoming a much more difficult problem to assess today than when the building industry was mainly craft-based and tendering reflected little more than the contractors' profit and their management skill in organising the output of their craftsmen to complete the job. In such a situation there was little opportunity for contractors to vary the deployment of their resources. With limited plant and machinery available and with production knowledge also limited to the best way to organise gangs of craftsmen to carry out the work, it was possible for the contractor to be given the quantities and for the lowest tender then to represent the lowest production resources.

Today there is still much work that comes into this category, but there are many elements in a modern building where different considerations apply.

Contractor's contribution to design and contract programme

There are occasions today when the contractor has a contribution to make which affects the design, speeds up construction, uses less production resources and, thereafter, produces a more economical job. We do not wish to overstate the importance of this contribution because a substantial majority of building works are, and will continue to be, small in value using fairly traditional methods. However, it is a factor not to be ignored and it is interesting that where sub-contractors are concerned, there is almost an undue readiness to consider the sub-contractor's contribution, whilst the main contractor's is often accepted with reluctance. This dichotomy is partly due to the fact that sub-contractors tend to represent specialist activities entailing considerable design responsibility, whilst main contractors frequently show reluctance to accept any such responsibility.

Engineering works differ somewhat from building works in that the reponsibility for detailed design is more frequently left in the hands of the contractor.

Cost planning techniques assist architects in designing economically. However, it is not always possible for the quantity surveyor or other members of the design team to know the best way in which the contractor's resources can be used to improve productivity in terms of cost and speed, particularly where a proprietary system is involved. For example, where a complete structure is fully designed before the contractor is chosen, the use of a proprietary system might well be excluded, despite its obvious economy, because of the subsequent amendments that might have to be made to the design in terms of detailed fixings and finishes. It therefore becomes necessary to consider at the pre-tender stage whether a contractor has a contribution to make or not, and if he has, when this should be introduced. This matter is discussed in detail in Chapter 4.

Production cost savings

The economics of the construction industry's production methods are clearly a most relevant factor in the cost of building and obtaining value for money for the industry's clients. In terms of a particular project, this comes down to the economic use of the production resources held by a particular contractor. Production resources are dynamic, not static; production savings are continuously being made in industry, and the building industry is no exception.

Construction has special problems in relation to increased productivity where design is separate from production. Most industrial firms have a great incentive to make production savings as the firms themselves benefit. In the construction industry this incentive is frequently limited by the design and is, therefore, confined to more productive ways of carrying out the design. It is difficult to promote savings where the design has to be altered to make such savings. Responsibility still remains with the designer, who is naturally reluctant to make amendments putting him at greater risk. Unless special contractual arrangements are devised, the contractor will not have the incentive to look for productivity savings where they would affect the design. Therefore, the opportunity to make production cost savings and the incentive to do so may be an essential ingredient of tendering procedures and contractual arrangements.

Continuity

Continuity of activity is perhaps one of the most important ways in which production and management resources can be used economically and, provided the tendering procedures are right, the client can benefit. Building has a unique problem in relation to the provision of continuity in that each building is on a different site and frequently user and client requirements will vary as well. Nevertheless a great many buildings are for clients who will want similar buildings on similar sites and, furthermore, at least some of the detail in dissimilar buildings will be the same.

Everyone takes longer to carry out a task the first time. Big savings arise the second time, rather less the third and so on until, after carrying out the activity many times, virtually no further savings can be achieved. A succession of one-off situations in management and production is bound to be uneconomical, and this is one of the biggest single areas in construction where better use of production resources can be obtained with careful planning.

This factor has a very substantial effect on tendering procedures, for, if continuity from a particular contractor is to be achieved, then it must be on a basis other than normal competition. This may make it necessary to plan the contractual arrangements for an initial project with a view to further projects being carried out by that contractor. Obviously, productivity savings arising from such continuity must be translated into a corresponding benefit to the client. Such methods as continuation contracts, serial contracting and term contracts are methods of doing this and are dealt with in later chapters.

Risk

Risk is also an important factor to consider in tendering procedure and contractual arrangements. It is sometimes thought that provided the contractor takes all the risk, accountability is satisfied. Frequently, however, this satisfaction is delusive, the true situation being hidden.

We define risk as the possible loss, which has to be stood by someone, resulting from the difference between what was anticipated and what finally happened. It is important to realise that a building owner takes on a considerable risk merely by the act of commissioning a building at all. The contractor never takes the total risk. The client, for example, takes the risk in relation to such matters as what the authorities will permit him to build on the site. The design team take the risk for the work they do, and so on. The question to be decided in the tendering procedure and contractual arrangement is how much risk it is appropriate for the contractor to take.

It does not always pay the building owner to ask the contractor to take the risk. Insurance companies, whose very purpose is to take the risk for you, charge a premium. There is no doubt that in every contractor's tender there is a hidden premium charging the client for the risk he has been asked to take, and in some cases, where the risk is very high, it will not be worthwhile for the client to pass the risk to the contractor, just as in some cases it is not worthwhile insuring against the risk.

It is significant that Government does not pay premiums to insurance companies to take the risk of destruction of Government buildings by fire. This is good practice in terms of accountability for public money. Such are the resources of Government, there is no need to pay others to take the risk. On the other hand, for small organisations the destruction of a substantial asset can be disastrous and they must pay someone else to cover the risk for them.

In building, risk is much affected by the state of the client's own resources. It would be wrong to push the analogy with insurance too far. It is often difficult to distinguish the risk that is appropriate for the contractor to take as an inherent part of his production activities, from the risk inherent in the construction itself. It is, of course, entirely appropriate that the contractor should take the risk for activities which are entirely under his control. It is appropriate, too, for the contractor to take the risk in operations which are basically under the control of the building owner or his designers where the building owner, because of his own circumstances, wishes the contractor to include in his tender a premium for taking that risk.

To illustrate this point, take on the one hand the conversion of an existing building where considerable risk is involved owing to the nature of the work. If the risk is too great, it may be impossible to get the contractor to tender at all. If it is moderate, then it would probably pay a client with substantial financial resources to take the risk himself and let the contract on a cost reimbursement basis. On the other hand, an individual of limited means, carrying out a conversion of his own house, may feel it is essential to get the contractor to take the risk and, therefore, get a fixed price.

A very common risk area today is that of fluctuations in prices of labour and materials, with the question as to which party should bear the risk involved.

Risk is one of the fundamental factors to consider in selecting suitable tendering procedures and contractual arrangements. It is a factor which depends on the client's circumstances as well as on the nature of the risk.

Conclusion

For convenience, we set out below what we consider to be the five fundamental factors to be considered in the selection of a contractor and the type of contract to be used. To some extent they overlap and they certainly interrelate.

(1)   the economic use of building resources
(2)   the assessment of the contractor's contribution in relation to the design and speed of construction
(3)   the incentives to make production cost savings and their control
(4)   continuity of work in all aspects
(5)   risk and the assessment of who should take it.

In addition it is worth noting the distinction between the selection of the contractor and the determination of the contract price. The traditional way of selecting a contractor, that of full drawings coupled with a bill of quantities, describing and quantifying the work to be done, results in the selection of the contractor simultaneously with his price being known. In this situation, the drawings and bills of quantities are tendering documents and also rank as contract documents. Selection and determination of price occur together.

In many other situations this is not possible. What in fact happens is that the selection of the contractor is made, but only the basis on which the contract price is to be obtained is agreed. The final determination of that price may take place many months later. Indeed, it does not have to be determined before the contract is made. If this point is appreciated, some of the more unusual methods of tendering will become easier to understand.

Finally, we emphasise that although all these factors should be considered for every project or group of projects, it is recognised that in a great many small projects, particularly those which are completely one-off and not part of a larger general programme, some of them will hardly apply. Tendering procedure must be considered in the context of the national scene. Most building contractors are relatively small and the majority of builders would be unable and unwilling to tackle a job of any size. Many small contractors, while making a first-class job of constructing a building to an architect's design, have neither the wish nor the ability to make the kind of contribution which a few national or specialist firms might make in some of the ways identified in this and subsequent chapters.

Having reviewed some of the wider issues affecting tendering procedures and contractual arrangements, we must not lose sight of the problem as it is so often seen by the client, namely, resolving the problems of COST, TIME and QUALITY – what could be called the 'eternal triangle' (see Figure 1) – and obtaining value for money.

Figure 1. Altering one element will have an effect on the others

These three elements may be held in balance or particular circumstances may dictate that one must take precedence over the other two.

Altering one element of the triangle will have an effect on the others, but, by evaluating the factors discussed in this and subsequent chapters, it is anticipated that the design team will be able to devise a tendering procedure and a contractual arrangement that will obtain the best value for money in the circumstances.

'. . . the relationship between agent and client is direct . . .'

Chapter 3

Accountability

In Chapter 2 we considered five main factors in selecting a contractor and formulating arrangements for a satisfactory contract. However, there is a further factor which can have a profound effect on these matters, namely accountability. Such is the special nature of this factor that it deserves a chapter on its own.

Accountability arises whenever one party carries out an activity on behalf of another. The agent must account to the principal for the actions he takes. The extent of that account (or level of accountability) must depend upon the principal's original brief and the degree of authority and responsibility delegated. In building procurement that authority and responsibility would be within a framework of principles, but the main priority may be cost, quality or time.

Background

Historically, the emphasis has been on accountability within the public sector although its relevance should be no less important within the private sector. The difference is purely one of reporting lines. In the private sector the relationship between agent and client is direct; any matters can be settled between the two parties. In the public sector the responsibility to the client – the public – is through an intermediary, usually the elected body, advised by an independent audit body. There is no direct relationship between the agent and the client.

The accountability of the professional advisor is an important aspect of the tendering procedure in construction because it is seldom possible for the building owner to see what he is buying before it is built. Even when this may be possible to some extent, for example in the case of a standard building, each will be on a different site and thus becomes unique.

Accountability differs from the other factors concerned with the tendering process, all of which are technical aspects requiring professional advice to determine their effect. It is a matter outside the strict building process since it applies wherever the client's agent is acting on his behalf in respect of purchasing, e.g. buildings, vehicles or stationery etc. Clearly therefore it is a proper matter on which the employer can be expected to have an opinion and it should be considered to be part of his brief, either directly or implicitly. In either event the professional advisors are responsible and accountable to the employer for both their actions and performance as agents.

The documentation for construction work is complex and technical and the amounts of money involved are nearly always large. This means that money can easily be misappropriated and there is more opportunity for corruption than in many activities although, in fact, the record is remarkably clean in the United Kingdom.

The client will wish to have an assurance that he is paying the proper price for the construction even though it may not be possible to determine the final price, or indeed the final extent of the construction, at the time of the signing of the contract. Where tenders are sought on the basis of competition, based on full documentation, the lowest price is normally accepted as being the proper price. Where a contractor is selected on a basis other than that of competition it can be more difficult to satisfy a client that he is indeed paying the proper price.

When it is decided to appoint a contractor on a basis other than equal competition, e.g. by negotiation, the reasons for making this decision may be complex and elusive. In many cases it will not always be possible to show clearly that there will be a monetary saving against competitive tendering. It must always be the objective to achieve value for money. There may well be times when the use of competitive tendering cannot achieve this and, particularly in the public sector, whether Central Government, Local Government or Government controlled organisations, it can be difficult to make a convincing case when accounting for the decisions thus made. Generally speaking Government, in its widest sense, will always seek to obtain tenders on the basis of competition as there is always the danger, however slight, that opinions leading to the selection of a suitable contractor by any other method may be subjective, and possibly corrupt, rather than objective.

The modern concept of public accountability

The concept of public accountability as we understand it today dates back to the Middle Ages. At that time production was craft orientated and sophisticated methods of modern production and construction techniques were unknown. Problems related to production were insignificant. If the State voted to spend some money on ships for its navy, the problem uppermost in its mind was to ensure that the money voted for the ships was actually paid to the shipbuilder rather than finding its way into somebody else's pocket. The method used in those days for the placing of orders was probably simple bargaining, implicit in which was the objective of value for money. The simplicity of the whole operation served to highlight the only points where the matter could go seriously wrong; these related entirely to graft and corruption.

As modern methods of production and construction technology have developed, it has become more important to be able to show that value for money has been obtained. While the standard of honesty of public officials, at any rate in this country, is high, the constraints imposed upon them to ensure that there is no corruption tend to work against the concept of value for money. Indeed, it is now possible for all the strictest canons of public accountability to be adhered to and yet extremely poor value for money to be obtained due to the waste of resources inherent in the present public accountability procedures.

The main reason for this is that, with complex technological production methods, the inefficient use, and therefore waste, of resources easily arises. The amounts that can be wasted in this way can well exceed those which in historic times found their way into officials' and favourites' pockets. To avoid such waste positive steps must be taken to ensure optimum use of contemporary production techniques.

Value for money

Value for money, as a concept, may be defined as achieving the optimum use of resources – resources comprising money, manpower, time and materials, each of which may be regarded as equally important. Until comparatively recently money was regarded as the most important resource in the public sector, the cheapest method of construction being the one most usually selected. With money becoming more 'expensive' the need to obtain more for a given amount has become imperative, and this has tended to emphasise the equal importance of constituent resources. For example, if a contractor is overpaid £1000 on a contract it is said that there is a loss of £1000; if there is a loss of £1000 worth of building manpower and materials through design errors then, again, there is a loss of £1000. On the one hand the money represents only financial resources which are no longer available for beneficial use, but on the other hand the money represents not only a financial loss, but also a loss of physical resources which have been used, wasted and lost.

Within both the public and the private sectors the overpayment loss will be obvious; the loss of manpower and material resources may not. In either case the professional advisors will be accountable to the building owner, to whom they are responsible. Government is concerned with the economic use of resources by the country as a whole. More resources used in building each project mean fewer projects. While Government is also concerned with the waste inherent in overpayment, this is very much the lesser of the two evils. Unfortunately, all historic tradition brings the overpayment to the fore, while the loss of resources remains obscure. Politically, too, overpayment is an easy matter to grasp and is highlighted in political debates as an example of the Government's ineptitude, while the loss of resources is difficult to identify and for this reason is more often forgotten or ignored.

A further factor in relation to public accountability arises from inflation. The argument has sometimes been advanced that the tendering procedure should be arranged so as to bring forward the commencement of the building in order that the earlier price level may be obtained. Once again, this may be a real saving to a private client depending on his circumstances and the use to which his money is put in the meantime. With Government, however, whose concern is the percentage of resources to be devoted to building, there is no saving at all so far as the internal inflationary situation is concerned. However, this needs to be balanced against other factors obtaining at the time; public finance accounting procedures and the way in which Central Government organises its annual financial allocations from time to time, may well place a premium on early financial year starts within the public sector.

The problem, particularly with public building, is to identify whether or not there will be a saving in resources in relation to the way in which the contractor is selected. This is a complex and difficult task, but in the long run the advantages to be obtained from many of the situations referred to in this book will not be forthcoming unless special attempt is made.

Numbers tendering

We deal here briefly with the matter of the number of contractors tendering, rather than elsewhere, because it is an aspect of accountability. In *Pre-Contract Practice* we advocate that, where competition is required, selective tendering should be used rather than open tendering, and a limit should be placed on the number of contractors tendering in accordance with the recommendations of the Code of Procedure for Single Stage Selective Tendering.

The arguments here are simple. If thirty contractors rather than five are persuaded to tender, the client might get a lower price on a particular job. On the other hand, if thirty contractors have to tender on all contracts there would be an enormous waste of estimating and managerial resources. Each contract would have to bear the burden of the tendering expense for an average of twenty-nine unsuccessful ones. It is salutary to consider that even with a limited list containing only five contractors, the successful tenderer has on average to bear the cost of four unsuccessful ones.

Conclusion

Accountability is an extremely important matter in tendering procedures and contractual arrangements. It is essential that it is looked at in terms of the resources used as well as the price paid. Due to historic traditions relating to public accountability, it is most necessary for the client's professional advisors to identify the effect that the procedure for the selection of the contractor may have on the use of building resources.

'. . . early selection . . .'

Chapter 4

Early Selection:
Two-Stage Tendering

In Chapter 2 we indicated various factors which should be taken into consideration in selecting a contractor. Some of these relate to the need to select a contractor early, prior to completion of the working drawings and bills of quantities. In this chapter we consider the occasions when early selection may be necessary and how it should be achieved.

When to select early

Although early selection has been defined as selection any time before complete documentation is ready, in practice it has advantages only when done at a very much earlier stage. Selection after completion of documentation in the normal manner is preferable unless those advocating early selection can truly justify its use. Basically there are two main reasons for doing so:

- to allow the contractor's participation in the design process
- to allow the contract to start and therefore complete sooner.

Early selection to allow the contractor's participation in the design process

The contractor's contribution to the normal pre-contract process can improve the quality of the end product and might be recommended for a number of reasons:

(1) *Technical contribution to design*
    Firstly it is essential to make an objective assessment of the value of the contractor's contribution, whether it will really amount to something positive which cannot otherwise be achieved, or whether it is more apparent than real. For

example, if it were intended to build several identical industrial units, the use of a proprietary cladding system might be justified. In such a case it would be unwise to process the design through to final production without having carried out a preliminary cost appraisal and ensured that the system meets the architect's requirements. Then an objective assessment must be made to determine whether such a system can produce a more economical or acceptable solution than, say, the use of a more routine non-specialist cladding material.

(2) *Management skills and buildability*
The contractor's input, at the design stage, in terms of management and practical skills, can often be of benefit in avoiding construction difficulties on site and assuring buildability.

(3) *Shortage of materials or time*
Once a contractor is appointed he is able to use his expertise in procurement to place orders and obtain materials which may be in short supply or for which there are long delivery times. This helps to minimise the risk of delay during construction. Additionally, being involved in the design process he can become more aware of the designer's intention and offer alternative construction solutions which save time on site.

(4) *Design alterations to suit specialist plant*
Prior knowledge of the contractor's expertise with particular plant may make it possible to design in such a manner as to take advantage of its use, and so speed construction; for example, in the use of 'slipform' techniques in the construction of the concrete core of a multi-storey office block.

Early selection to achieve an earlier commencement and completion

The client may wish to start work on the site before production information is fully available, to achieve:

(1) *Early commencement*
A subsidy or grant may be directly linked to a statutory date for starting work on site. Subject to annual accounting, projects are sometimes only sanctioned if a start can be

made before the end of the financial year. In these cases an evaluation has to be made between the extra costs involved in a start before planning is complete, and the benefit of a grant. Early selection and parallel working (see below) may be preferable to an inefficient competition based on inadequate information.

(2) *Early completion*
The sooner the start, the sooner the completion, provided of course design work and construction are programmed and integrated properly. A short overall programme of design and construction also assists in reducing the client's interest payments on funding and site costs. In this case the contractor will not necessarily be participating in the manner described above, but could be working in parallel with the designers. For example, the contractor might be building the foundations while the final details of the structure are being designed, or erecting the structure while the details of the finishings are being finalised.

Such a system has its dangers; hence the need for the most detailed programming. It may be even more expensive in terms of construction, and some discrepancies between work in hand and work still to be detailed may perhaps arise. However, the income derived from the building investment may well make it worthwhile to save even only a few weeks. Income gained could be used to help fund later phases of the scheme which could perhaps follow a more traditional procurement route. Once again, an objective financial assessment must be made of the possible extra cost resulting from improved speed and this must be weighed against the increased revenue.

How to select: two-stage tendering

It is possible in any of the above cases to select a contractor and carry out a single negotiation on the basis of cost alone. Some of the occasions when this might be appropriate are described later, for example in Chapter 6, Competition and Negotiation. Whilst this may satisfy the requirements for speed and quality it can result in increased risk to the client with respect to the finally negotiated price.

On the other hand the process of two-stage tendering seeks to balance the cost-time-quality triangle, and can be used where early selection is required but where there is no need to negotiate with a single contractor, thus avoiding some of the risk inherent in negotiating with one specific contractor.

*   The term two-stage tendering is used to describe the procedure where a contractor is selected in one operation and the contract sum is agreed in a second operation.

The first stage, the selection of the contractor, is conducted on a competitive basis in the pricing of brief but precise documents related to a preliminary design. This will provide the basis of pricing for use in subsequent negotiation, but not a contract sum.

In some instances the first stage may be preceded by a preliminary stage which involves interviewing selected contractors to establish the resources they have and the contribution they can make. It is undesirable that this preliminary stage should become the end of the matter. It can only be a subjective assessment leading to the first stage tender.

The NJCC in their Code of Procedure for Two-Stage Selective Tendering state that:

In order to minimise the cost of tendering overheads and to facilitate the selection of a contractor at an early date, documentation for the first stage should be kept to a minimum consistent with the need to:

*   provide a competitive basis for selection;
*   establish the principles of layout and design;
*   provide unambiguous pricing documents related to preliminary design and specification information in forms sufficiently flexible to be suitable as a basis for pricing the second stage tender. Provision should be made for adjustment of price fluctuations during the period between first and second stage tenders;
*   define the obligations and conditions of, and state the programme for, the second stage procedures;
*   state the Conditions of Contract.

Stage one: documentation and tender

Three methods of preparation of the documentation for the initial selection are set out below. However, there are many others and it is up to the design team to choose the most appropriate, bearing in mind the client's requirements and the particular circumstances. The three methods briefly described are:

(1)   The general estimating method

(2)   The approximate quantities method
(3)   The cost plan/proprietary design method.

(1) The general estimating method

This approach is most suitable for large contracts of a general nature where no particular design contribution is called for from the contractor. The tendering documents for selection of the contractor should include:

(a)   A preliminaries section giving particulars of all site management costs and head office overheads and profits etc. This can normally be priced on a fixed price basis provided the general scope of the project is known. It need only be adjusted later if the scope of the scheme is materially altered.
(b)   The contractor's allowance for head office overheads and profit in relation to:

   ●   his own work
   ●   his own sub-contractors' work
   ●   nominated sub-contractors' work.

(c)   The basis of labour rates. This can either be in terms of a calculation of an all-in labour rate, or alternatively of a net labour rate with some of the indirect labour costs calculated as a total sum to be allowed against a total calculation of the labour required for the job.
(d)   Some means whereby the productivity of labour can be assessed. This may be a breakdown of pricing for typical bill of quantities items, with notional quantities relating to work the contractor is likely to carry out himself.
(e)   The method by which sub-contractors' work will be dealt with.
(f)   The method by which materials will be dealt with, making allowances for waste in typical cases.

Refinements can be made to the above list and details can be asked for in tendering documents for any particular items likely to have an important effect on the cost of the contract.

(2) The approximate quantities method

This method forms an alternative to the general estimating method and may be used in similar circumstances. However, whereas the general estimating method establishes labour constants and material prices, the approximate quantities approach establishes competitive prices for measured rates.

The first stage documentation will comprise a bill of approximate quantities, including the specification, materials and items which are likely to occur in the final design. The quantities stated would be approximate in that they had been estimated from the early design details. All the sections of a normal bill of quantities would be included, e.g. preliminaries, preambles, measured works etc.

After a contractor has been selected and the design completed, the approximate quantities can be confirmed and the contract price established. It is important that the initial quantities are as near to the 'designed' quantities as possible so that the level of overheads, plant and profit originally included by the contractor does not have to be re-negotiated or adjusted.

(3)   *The cost plan/proprietary design method*

This method is normally used when the contractor has a fundamental contribution to make to the design through the use of a proprietary system. The object is to select a contractor whose proprietary system best meets the requirements of the architectural scheme, and to judge the offers received in terms of best value for money.

With this method it is important to have preliminary interviews with the contractors to establish which have systems that can be adapted to suit the architect's sketch plan. There is no point in asking a contractor to tender if he has to alter his system uneconomically or if the design has to be restricted by the limitations of his system. A number of interviews with possible contractors may be necessary to select a short-list of tenderers.

In preparing tendering documents it is essential to identify the areas of work under the following six main headings:

(1)   the works to be carried out in a traditional manner, such as the external work, external services and possibly the foundations
(2)   the work included in the proprietary system
(3)   the nominated or approved sub-contract work
(4)   the fixing or assembly of suppliers' work in the superstructures

(5)   preliminaries and site management
(6)   head office overheads and profit.

By the very nature of this work it is not possible to identify these areas accurately. Some proprietary systems take in more work than others. The tendering documents must provide for flexibility so that the contractor who can economically incorporate a certain piece of the work in his proprietary system will be able to do so. However, the architect, quite rightly, will wish to have as much control as possible over those items which are not an essential part of the proprietary system. The following documentation is appropriate:

- A bill of quantities or approximate quantities for the external services and external works. The actual limit of measurement will depend on the circumstances and each case must be judged on its merits. The external elements are normally unaffected by whichever proprietary system is adopted and the architect is free to specify these as he chooses.
- Bills of quantities or approximate quantities for the elements unconnected with the proprietary sysem, e.g. foundations. Care is needed because in some cases these elements can be designed independently of the system, but this is not always so.
- A performance specification for the work covered by the system should be prepared. This must have built-in flexibility to allow for any acceptable variation between each system. Wherever possible, provision must also be made for a breakdown of the price for the system content – this is for control purposes once the contractor has been selected. If included, quantities must be on a functional basis, e.g. areas of floor related to superimposed load.
- The remaining work will largely be covered by sub-contractors who will either be nominated or agreed and approved between the architect and the main contractor. This area of work is, for tendering purposes, best covered by giving details of the cost plan quantities with a specification for each element. The contractor can then see what is required and if his proprietary system encroaches on this work. This is important because it allows the price for the area of building covered by the proprietary system to be seen in the context of the building as a whole.
- There will be certain items of builder's work in the superstruc-

ture, such as the fixing of doors and fittings, which can be covered either by a schedule of rates or alternatively by approximate all-in quantities.

- Site management costs and head office overheads and profits will be dealt with as in the general estimating method.

Evaluation and selection

Whatever the format of the first stage, when an offer is received it will not appear as a straightforward tender figure on which an immediate comparison can be made with the other tenders. An objective evaluation has to be made of the various tenders, weighting items in order to arrive at a common denominator. For example, a contractor who has decided to use two tower cranes instead of one will have included a greater amount for this item but may be saving labour.

Similarly, whilst one proprietary system may be more expensive than another, its effect on the remainder of the building may result in an overall saving. Conflict may arise between the objective assessment in financial terms and a subjective opinion on performance or aesthetic grounds. When this occurs it is essential not to mix the assessments. Those matters which can reasonably be evaluated in terms of money should be assessed in that way. But overall, the cheapest tender may not provide the best solution. In such cases evaluation is difficult and however objectively carried out, will be bound to contain some element of subjective assessment.

In making the evaluation it is worth pointing out to the client the extent of the area of opinion as against straightforward objective assessment, and the possible effect this could have on the selection. Additionally, once they are accepted, the client has to pay the level of overheads and profit irrespective of the amount of work to be undertaken. Therefore they carry as much, if not more, influence and weighting than the rates and pricing in the first stage tender.

Stage two: determination of the contract price

Having evaluated the offers and advised on selection it is necessary to proceed to the second stage, the determination of the contract price. This should be done as early as possible but must be tied to the evolution of the design. Where a proprietary system is involved there could be a considerable amount of design refinement to tailor the system to the architectural requirements. During this period of evolvement, negotiations on the price within the context of the tender will be carried out and finally a firm contract price will be agreed reflecting the finalised contract drawings and documents.

The second stage documentation format should closely follow that used for the first stage. The whole second stage document must be in a form to which cost control can continue to be applied during the construction stage. The second stage document will be priced on the basis of the first stage pricing data. If the programme is urgent, entailing parallel working, it may be necessary to agree the pricing on stage two progressively.

To maintain the time scale a compromise over the quality of the final documents or a premium on the contract price may be unavoidable. This can occur if provisional allowances have to be included where final details are not known or negotiations cannot be completed.

Advantages and disadvantages

In the appropriate circumstances two-stage tendering has an effective part to play in achieving a satisfactory tender price. However, there are both advantages and disadvantages to its use which must be made clear to the client before it is adopted.

Advantages

- An overall tender price is usually known before work commences.
- There is a saving of time over single stage tendering methods due to the tender/design overlap prior to a site start.
- Control of the design and workmanship on site is maintained by the client's consultants.
- The procedure still enables full tender competition of subcontract works to be carried out as a parallel operation.
- Fewer disputes should arise as the contractor has been more involved with the design and document preparation and has a better understanding of the requirements.

Disadvantages

- The contract sum may be more expensive than the single stage method because of the negotiated element and reduced level of competition.
- Delays in providing design information can result in project overruns and additional costs.
- Second stage tender negotiations could prove difficult and protracted. In such circumstances a higher than market price may have to be accepted if delays are to be avoided while alternative tenders are obtained.

'. . . someone from a different profession . . .'

Chapter 5

Drawings, Specifications and Quantities

In Chapter 1, in the section on contract conditions, a number of different contracts commonly in use were listed and later in the book specific formats are looked at in depth.

One of the basic criteria for determining the type of contract best suited to the client's needs is the information that is available, or that can be made available, for tendering in the time scale dictated by the client's brief. In Chapter 8 there is a tabulation of the various possibilities in relation to fixed price contracts.

Traditionally, information to enable a contractor to construct a project is set out in drawings (normally taken also to include Schedules) and specifications, with the administrative details relative to the contract being set out in bills of quantities. All are essential as part of the tender documents in traditional procedures.

Good practice in the production of drawings and specifications in the context of traditional working relationships has been set out in detail in *Pre-Contract Practice* and there is no need to repeat that information here. What is required, however, is an appreciation of the significance of the various documents and their contents in the non-traditional contractual arrangements.

Non-traditional arrangements

When the traditional mould is broken the same information is still required to complete the project, but it may have to be provided in a different format.

One of the most significant factors to affect the documentation is the decision on whether or not the work should be measured in a bill of quantities for the contractor in the tendering documentation. If the work is not to be quantified, the written part of the contract documentation must be provided by other means. Usually the specification becomes a contract document in its own right (either as a traditional specification or a performance specification) and this document, or a further document, then has to contain the administrative details usually contained in the bills of quantities.

The decision whether or not to quantify may be evaluated in relation to the criteria set out below.

Criteria for quantification

In our opinion, quantification should be accepted as the rule because of the major advantages set out below. Only where it can be shown positively that there is a case for not having quantities should quantification be dispensed with.

The advantages of the quantities system are:

- With quantities the estimating risk is considerably lower for the contractor.
- With quantities provided, the competition is much fairer as all contractors are tendering on the same basis.
- It is more economical for the contractors tendering, and therefore for the whole building industry, thus lowering overheads.
- The quantities system automatically allows the total price to be analysed in great detail, thus providing cost feedback on the job which, in turn, can be used statistically to help in cost planning other work.
- Quantities provide one of the best means of cost control for variations on the contract.
- Although at present in neither the Standard Form of Contract nor the Standard Method of Measurement is it envisaged that quantities will be used for management purposes, e.g. for ordering, they do in fact provide documents which are a great help on the site for this purpose. Furthermore, it is likely that new methods of measurement, coupled with computer application, will lead to their far greater use for management purposes in future.
- The process of quantification before tender is one of the severest tests as to whether what has been drawn and specified can in fact be built. Even in the best drawing offices, having someone from a different profession examine

the drawings and analyse the construction in detail is a substantial help in foreseeing problems not apparent at first sight.

There are, however, also some disadvantages:

- For the work to be quantified the design must be substantially complete. As mentioned in previous chapters, there may be situations where the contractor has a contribution to make to design. In such cases quantification must at least be amended, e.g. by procedure under two-stage tendering.
- For small jobs quantification can be tedious and expensive in relation to the size of job.
- Quantification that appears to be complete holds dangers if it is in fact partially speculative but sufficient to allow a tender to be obtained and a contract to be signed before design work has been adequately done. All too often, in both private and public sectors, the pressure to sign a contract by a certain date militates against proper completion of the detailed design work. The use of quantification as a means of overcoming this difficulty may place the contractor at contentious risk and result in dispute.

In our view the advantages of quantification are generally overwhelming. Where, however, particular circumstances make it disadvantageous, it is highly desirable for cost control purposes that at least an analysis be made of the price. If possible, quantification should then take place after tendering, in agreement with the contractor.

This emphasis on quantification is not to deny the importance of the specification. In terms of the quality of work to be built, specification is paramount. In terms, however, of proper financial arrangements for a contract, quantification of what is to be built is advantageous.

Bills of quantities

Pre-Contract Practice and *Contract Administration* deal with the normal situation with regard to bills of quantities. As far as tendering and contract are concerned, the principles are the same whatever layout is used for the bills. Trade layout is the easiest on which to estimate, while elemental bills are more difficult for estimating but provide a ready basis for cost analysis. Operational bills, because they are production oriented, should allow more accurate estimating and are the most useful for management purposes during the contract. Annotated bills may be any one of these three, having location and specification notes against the bill descriptions, which can be useful for management purposes as well.

The work to be executed should be fully and properly planned beforehand in order to get the full advantage of whichever type of bill is used. It is likely that for some considerable time this method will be used in the majority of cases because of the formidable advantages it offers, particularly in terms of financial control.

Approximate quantities

While accurate quantities should be used wherever possible, there are circumstances in which it can be shown positively that approximate quantities are necessary, such as:

- Where speed is of paramount importance and the general design established, it may be necessary to select a contractor before production drawings can be completed. In such cases, sufficient is known of the design for approximate quantities to be produced on which a tender and a sound contract can be based.
- For work in the ground the conditions may not be known sufficiently well for it to be practicable to produce accurate quantities until the work is carried out. Thus the perimeter of a building may be known but the various soft spots and general depth of foundations may have to be broadly assessed.
- In accurate bills of quantities there will be what are described as provisional quantities to be remeasured when the work is actually done. Thus accurate measurement of the foundations may be possible except for the underpinning of a particular wall, which has to be a provisional quantity until that wall has been opened up. Another example is the extra for excavation in rock, where the foundations can be measured accurately but the amount of rock to be excavated will not be known until the work is done.

Where bills of approximate quantities are used for tendering, the procedures will generally be exactly as with accurate bills; but the Standard Form of Contract for use with approximate quantities, prepared by the Joint Contracts Tribunal, should be used. The differences are, however, minor and the chief problem arises in the post-contract administration, particularly with regard to cost control. There are three main choices available at this stage:

(1) The work can be remeasured at the end of the contract and this then becomes the basis for the final account. In this case, cost control and interim valuations become more difficult and a good deal of interim approximate measurement is necessary if these functions are to be properly performed.

(2) The work can be measured as soon as the production drawings are available, and substitution bills prepared in place of the approximate bills for the administration of the contract. This is the best method for control purposes but undoubtedly means more work as there are then likely to be further variations on the production drawings which, in turn, entail a final account having to be measured from the substitution bills.

(3) Particularly in large contracts where there is a resident quantity surveyor, the work may be measured immediately it is executed. This keeps the measurement up-to-date but it is of more limited value in forecasting future expenditure.

Bills of approximate quantities are obviously less useful than accurate bills and generally cannot be used for management purposes. In such cases there is little point in annotating them, as the work that they describe is to some extent hypothetical.

Schedule of rates

In schedule of rates contracts the quantities are measured afterwards and the tender is, therefore, only on a unit basis. It is obvious that a schedule of rates is unsuitable for a contract where the quantity of work is likely to affect the rate. It is most suitable for jobbing work or redecoration. We describe one application of schedules of rates under term contracts in Chapter 13, but they have other applications.

Schedule of rates tendering can be competitive but obviously it is difficult to make an objective evaluation without quantities. Frequently, therefore, schedule of rates tendering is done on the .

basis of a pre-priced schedule allowing the contractor to add or deduct a percentage with respect to each particular trade. Such a system of tendering and contract is common in some parts of Europe. Cost control in the normal sense is impossible, but this is not to say that the work cannot be properly budgeted by putting predicted costs on the project before it is started. Furthermore, the costs can be accurately accounted for. In order that valuations may be accurate and up to date, the work must be measured regularly.

Specification-based documentation

There are certain cases in which it may be more appropriate to provide the documentation in the form of drawings and a specification or performance specification. Subject to certain qualifications made later with regard to the performance specification, we do not recommend that a specification be the only tendering document for work of any magnitude. However, for small works it may well prove to be appropriate by reason of its simplicity and direct relevance to the work to be done. This is particularly so where small-scale alterations and conversion work are being carried out, or in maintenance contracts.

Normal specification-based contracts

A specification for such a project has three purposes:

(1) It will be the tendering document and, therefore, must be put in a form which allows the estimator to price it easily. This means that trades must be separated where possible, as this is likely to be the basis of the contractor's arrangements with his own sub-contractors.
(2) It will, on acceptance of the tender, become a contract document and therefore must stand up without ambiguity as part of a legal contract.
(3) It will be a management document. On the one hand it will tell the contractor exactly what work is physically to be carried out on the site. On the other hand it will serve the architect or quantity surveyor as a cost control document so that valuations for certificates and variations can be properly ascertained.

Some quantification is still possible and desirable, but will not have the same status as when a bill of quantities and a Standard Form of Contract with quantities are being used.

Performance specifications

Performance specifications may well be used on jobs of any size and may be combined with, or part of, bills of quantities according to the nature of the project.

The performance specification specifies the performance required for the particular section of work to which it relates and the contractor is permitted to build in materials of his choice and to his design provided only that the defined performance is achieved.

If a performance specification is used for competitive tendering it is likely that the various contractors will come up with different proposals, and these have to be evaluated. Great care has to be taken in drafting the requirements of the performance specification as, if these are too rigid, little is left to the initiative of the contractor. Alternatively, if they are too loose the evaluation becomes more difficult as the answers will vary more widely. Economic criteria must be taken into account when formulating the specification, otherwise the cost will be too high and the specifications will have to be amended by negotiation if the whole exercise is not to be abortive.

Although a performance specification does not include quantification in the same way as bills of quantities for independently designed work, nevertheless it is important to quantify it as far as possible for the purpose of cost control. Whereas independently designed work can be measured in terms of the work to be carried out, in the case of performance specifications the measurement has to be in terms of the function of the specification. Thus, an independently designed suspended floor will be measured showing the concrete reinforcement and formwork together with the beams holding it up, but the floor defined in a performance specification will have the quantification in terms of the total floor area to be suspended. No separate measurement can be made for the beams supporting it, although there will be separate quantities whenever the performance requirements change or wherever the physical requirements of the building clearly indicate a change in the construction of the floor.

It is important that cost control should be available to the client even though performance specifications are used. The best way to ensure this is to separate the various sections of work into the smallest convenient functions so that the price for each section is isolated. In this way a substantial variation to a particular section can be more easily identified later. Thus, if the electrical installation is to be tendered for on the basis of a performance specification, it is desirable to keep the wiring for the rising main and intake separate from the distribution mains, and separate again from the wiring to outlet points in various sections of the building.

There is no reason why, following tenders on a performance specification and a particular contractor having been selected, the quantification for his particular solution to the design problem should not be made and incorporated in the contract documents for the purpose of cost control. This should be stated as a condition of tender so that there is no objection later.

'. . . *a network analysis should be prepared* . . .'

Chapter 6

Competition and Negotiation

In this chapter we consider five main aspects of competition and negotiation:

(1) competition
(2) the relationship between competition and negotiation
(3) when to negotiate
(4) the basis for negotiation
(5) how to negotiate.

Competition

A contract procured by competition will involve tenderers submitting quotations based upon documentation common to all. The documentation, as described in Chapter 5, will invite tenders which incorporate as priorities one or a number of the following:

- price
- time
- quality
- design
- management.

Quotations of price and time are definite and bear direct comparison with each submission. Interpretation of quotations for design, quality and management introduce opinion and judgement by the employer and his advisers before acceptance. The common understanding of competition relates to finance and usually is restricted to tenders where price and time form the major ingredients of the submission.

In such cases, the Code of Procedure for Single Stage Selective Tendering (1989), advising on the number of tenderers invited for the size of contract, should be adhered to in preparing a list of tenderers. Careful consideration should be given to the choice of tenderers, both in their ability to carry out the work in the time given or quoted and their financial stability. Early notification should be given to each tenderer of the pre–contract and projected contract programmes, and dates should be set and agreed well in advance for issue of the tender documents and receipt of tenders.

The relationship between competition and negotiation

Most contracts involve both competition and negotiation to a certain extent, and this might be represented by the formula:

$$C + N + X = 100\%$$

C represents the value of work in the tender submitted on a competitive basis

N represents that which is negotiated

X which is usually very small, represents those parts of the contract which are neither on a competitive basis nor negotiable, but subject to factors beyond the control of the parties to the contract (such as statutory authority charges or costs arising out of legislation).

It is possible to have a contract which is virtually 100 per cent competitive and it is also possible to have one which is nearly 100 per cent negotiated. It is, however, rare to achieve percentages this high. The majority of contracts are a mixture of both and it is worth noting that competitive contracts will frequently have the major part of the work negotiated, perhaps in the sub-contracts. Similarly, there is many a negotiated contract where the majority of the work is actually on a competitive basis. The terms, therefore, tend to be related to the method of selection of the main contractor rather than a definition of the proportion of work under competition or negotiation which makes up that contract.

The distinction between a competitive and a negotiated contract has nothing whatsoever to do with whether the contract is a fixed price, target cost or reimbursement form of contract. A fixed price contract might well have the contractor selected singly on a subjective basis and the price subsequently negotiated. Equally, a cost reimbursement contract, with a fixed fee for management, can be awarded on a competitive basis.

Negotiation

Negotiation falls under two headings:

(1) Pre-contract negotiations – based on estimated costs or sections of work to be let on future tender arrangements
(2) Post-contract negotiation – which will nearly always involve a large element of known costs and, if there are no comparable contract rates, the examination of invoices, daywork charges and the like submitted by the contractor in reclaiming his loss and expense on any variations.

When to negotiate

As a general principle, on grounds of accountability, one should select a contractor and negotiate only where it can be demonstrated to the client that there is an advantage in doing so. The following are situations in which the contractor might be selected by a method other than competition and in which it may be necessary to negotiate a price:

(1) Business relationship
The client may have a business relationship with the contractor: for example, reciprocal trading, where an employer will go out of his way to employ a contractor who trades with him even if it means single tender action and negotiation. Another example would be a joint partnership in a development company or wholly-owned subsidiary. Although in these cases it may be difficult to prove to the client the advantages of negotiation, nevertheless, taking his total interest into account, it may be desirable.

(2) Contractor-financed projects
If the client finds it difficult or impossible to finance the project in any other way, then it may be desirable to select a contractor who will be willing to finance the project in the short term. In this situation, however, it is essential that the cost of financing the project is separated from the normal construction costs.

(3) Continuation of contract
The client may have let an initial contract in competition and then find that another project with a similar design comes on programme, in which case it may be easy to prove that it would be more economical to let the second contract to the contractor

carrying out the first one. This option is dealt with in detail in Chapter 13.

Sometimes a client will have such confidence in a particular contractor, because of a large amount of work that he has carried out for him, that he will wish to select this particular contractor and negotiate, even though the particular project may be quite different from previous work. Such a course may be justified but should be carefully watched as there may be a temptation for the contractor to push up the price, until a completely uneconomic situation has arisen.

There can be a learning curve in this situation. Care should be taken to ensure that a correct balance is taken, making due allowances for both increases and savings in cost of management, supervision, purchase of materials, letting sub-contract and all site operations.

(4) *Special circumstances*
In particular areas there may be only one contractor available to do the work. In such a case it may well be sensible to accept the situation rather than import an outside contractor. An example would be fire or storm damage to a building during its construction. Loss adjusters and insurers become involved and in some cases the name of the client could change. However, the contractor on site would need to be instructed and negotiations are really the only solution.

(5) *Special expertise or equipment*
A certain contractor may be the only one available with either the expertise or the special equipment to carry out the particular project. Such a situation must be carefully assessed and if it is found that this really is the case, it is better to accept it and conduct a proper professional negotiation.

(6) *Tendering climate*
At times when the industry is grossly over-stretched, it may well pay to negotiate a particular contract. The alternative might be to receive in competition a price embodying a very heavy premium as the contractor did not particularly want the job. A contractor will often be prepared to negotiate a reasonable price at such times, because he places value on the goodwill of a particular client. Care must be exercised in the choice of contractor with whom to negotiate.

(7) *Quick start*
If a very rapid start is required, it may be necessary to negotiate.

This situation must be looked at carefully, however, as it is possible to select a contractor in competition in a very short time indeed. Where a contractor has gone into liquidation, it may be an advantage to negotiate with another contractor. In this case all the design work will have been completed and the savings of time may justify selecting a single contractor and negotiating.

Other financial considerations may be far greater than the difference between negotiation and competition. An example would be a high consequential loss situation arising from material damage to a building from fire.

Generally

The above list has given an indication of the situations where the total contract might be negotiated with a single contractor. There are many other cases, however, where the contractor is selected in competition but where, because of his early selection for some particular reason, a considerable amount of negotiation of the contract price will occur. However, it will be on the basis defined in the competitive bid, rather than a negotiation from scratch.

Finally, there is some negotiation in nearly every contract. For example, even on a fixed price traditional form of contract with bills of quantities, there will be some negotiation in settling the final account. Although the Standard Form of Contract gives the quantity surveyor the absolute right and duty to determine a fair price for the variation, nevertheless, in practice, this is frequently negotiated and agreed with the contractor.

Bases for negotiation

There are only two bases for negotiation:

(1) in relation to the competitive price for similar work under similar conditions in another or previous contract
(2) an agreed assessment of the estimated cost to which will be added a percentage for head office overheads and profit.

The first method has certain advantages in respect of accountability. For example, it can be explained to either shareholders or an overseeing Ministry that although the contract was negotiated, the negotiation was based on another contract which had been competitive. A 'nominated bill' may be produced which the contractor has used on another contract, not necessarily with the same client, and which would form the basis of the negotiations.

There are, however, great dangers in this method. The other job may not form a suitable basis for all sorts of reasons, some of which may not become apparent until after negotiation has gone quite a long way. If such a method is adopted, it is essential that the client's quantity surveyors know as much as the contractor about the circumstances of the other job and this, by its very nature, is difficult. For example, last-minute changes in the original competitive tender may have seriously upset some of the individual prices, although such distortions may not have upset the original contract, particularly if there were no variations on those items. However, on the new contract, with different quantities, an unfair total price may result.

Where possible, the evidence of actual costs from the final account of the previous contract should be examined and brought into negotiations. Particular attention should be given to obtaining up-to-date quotations for specialist items and nominated work. Also, where the contractor intends to sub-let trades, then where time allows, competitive tenders should be obtained from three or four firms. There is a time limit to the use of historical cost or price information. It is important to review all aspects of the pricing in an endeavour to obtain a fair result for both the contractor and the client.

The second method, that of assessing the estimated cost of the job, is logical because this is the normal method of tendering on a competitive basis. In this situation, the price is built up from the suppliers' and sub-contractors' quotations, together with an assessment of labour and plant requirements. Finally, after site management costs have been estimated, a percentage figure is added for head office overheads and profit. Both the client's and the contractor's quantity surveyors will participate in producing the price on this basis, which may well be documented in a normal bill of quantities.

The dangers in this method are that the estimate may be conservative and not so keen as in a competitive situation. Risks which the contractor would be prepared to take in a competitive situation, to get the job, may be exaggerated or indeed removed in the negotiated situation. The best way of getting over this is to use the assessment of estimated costs as the basis, but always to check each result in a competitive light.

How to negotiate

Whichever basis is decided upon for the negotiation, there are certain fundamental principles that should be established as a prerequisite to negotiation.

(1) *Equality of negotiators*

In negotiations it is both courteous and practical to ensure a reasonable equality of status between the parties and a fair balance of numbers. It may not be sensible of course to expect the managing director of a major contracting firm to attend a meeting with the principal of a small architectural practice and his quantity surveyor to discuss only a modest sized contract. In such a case the contractor will be represented by one or two seniors from the contract management and surveying departments. At no time should meetings be arranged in which the senior level on one side tries to obtain agreement from a lower level on the other. It is important that in managing negotiations, people's duties and delegated powers are understood and defined and meetings arranged that are balanced and therefore productive.

(2) *Parity of information*

This should be established at the outset. Much of the information will come from the contractor's sources but if he is not prepared to share this with the client's surveyor, suspicion will immediately result. The negotiation will take much longer and the client's surveyor will have to get his own information. One would immediately question the motives of a contractor who was not prepared freely to give the other side the information on which the estimate was being based. This applies particularly to competitive quotations for materials and plant.

The surveyor, on his side, must respect the complete confidentiality of information passed to him. There is no doubt that parity of information is the best basis on which to build up mutual trust between the two sides, and this is the most likely way to get a successful negotiation.

(3) *Structure of estimate*

The basis of negotiation must be decided from the outset. If it is to be a nominated bill, that bill must be analysed and broken down to the original basis on which the estimate was made. Only thus can variations relating to the new job become apparent. If, however, the estimate is built up from an assessment of the estimated costs, then the structure of the estimate must be agreed initially. Agreement should be reached at this stage on:

- where the head office overheads and profit are to be priced

- whether the plant is to be priced in the trades or in preliminaries
- how the labour on-costs are to be established
- the relationship between working and managing foremen
- firm price element
- risk.

Contractors have many different ways of structuring their estimates and there is no need to insist on any particular method, as long as it is clear how the estimate is structured so that each element can be identified properly.

(4)   *Apportionment of cost*

It is often useful to establish at an early stage an indication of where the cost is likely to lie on the project. An example might be as shown in the table and Figure 2.

If such a breakdown is made it will be seen, as in this particular example, that some 35 per cent of the contract is not to be negotiated at all at this stage. Furthermore, another 30 per cent will be covered by competitive quotations (materials and contractor's own sub-contractors). This, therefore, leaves some 27.5 per cent to be assessed and agreed, with 7.5 per cent (the head office overheads and profit) to be negotiated (see Figure 2). Different jobs will, of course, have different percentages; the purpose here is only to put the negotiation into its proper perspective.

Before examining the above items individually consideration should be given to the period of the contract, on which so many of the items depend. Ideally a network analysis should be prepared and the critical path found and examined to establish the time required and where savings in time can be effective. Consideration must be given to the optimum time for construction, always bearing in mind the development as a whole. As so many of the items relating to site management and plant are directly related to time, careful consideration must be given to this aspect.

Detailed consideration should be given to all materials and sub-trades where there is evidence of a prolonged period from order to site operation. For a negotiation to be successful it should be right in price, quality and time. Delays on the contract period can result in loss and expense claims which could materially invalidate the negotiated contract sum.

	%
Site management	7.5
Contractor's own labour	15
Direct materials	20
Plant	5
Contractor's own sub-contractors	10
Nominated sub-contractors	25
Nominated suppliers	5
Provisional sums and contingencies	5
Head office overheads and profit	7.5
Total	100

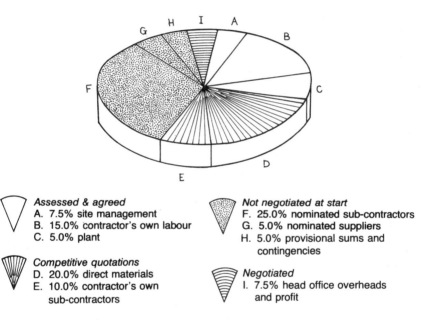

Figure 2 Apportionment of cost

(a) Site management
Having previously clearly defined what is to be included under
this heading, it is a matter of making a reasonable assessment of
what is going to be required. By far the biggest cost will be the
actual management staff and a detailed list should be drawn up
of the staff required. It is desirable here that reference should be
made to competitive estimates, and reasons given if the assessments of site staff on the negotiated project differ from a similar
competitive one. The surveyor will also have much information

from his own office and from other negotiated contracts and, whereas he should never disclose another contractor's figures, it is perfectly in order to question the figures of the particular contractor with whom he is negotiating, on the basis of his background experience.

(b) Contractor's own labour

This is the most difficult part of the assessment and the best check is to establish net prices for the work measured in position and then, by deducting the material content, to establish whether the labour is reasonable for each section of work. It must be remembered that an assessment is clearly needed for the total costs of the labour in order to establish the direct on-costs related to labour.

(c) Plant

This will either be priced in relation to the particular work it has to do (for example, excavation) or else it will be general plant such as tower cranes for which an estimate must be made of the period for which they are required. Small tools may come under this heading or may be expressed as a percentage of labour.

(d) Direct materials

These will, in most cases, be based on competitive quotations but an estimation has to be made of waste and the risk element: for example, faulty materials where the faults are disputed.

(e) Contractor's own sub-contractors

Usually these will be on the basis of competitive quotations but it may be necessary to take in subsidiary attendance in order to establish the complete cost here.

One problem that may arise concerns the amount of the sub-contractor's quotations, of which the lowest may still be considered too high. In a competitive contract the contractor will shop around after he has been awarded the contract, to get in lower quotations. In such a case it may be desirable to accept the lowest available quotation for the purpose of the tender, but to agree to substitute a lower quotation if one is later deemed advantageous so that the employer gets the benefit of the shopping around. If time allows, however, it is always better to get competitive quotations before finalising the contract.

(f) and (g) Nominated sub-contractors and nominated suppliers

Only attendance will have to be estimated here. It is assumed that the profit on them will be included in the overall profit for the job.

(h) Provisional sums and contingencies
These are part of the client's budget put into the estimate and therefore there is nothing to negotiate.

(i) Head office overheads and profit
These should be shown separately: the head office overheads, which ought to be a known factor with any contractor and therefore easily substantiated, and profit, which must provide for:

- the return on capital employed
- the risk involved in carrying out work at the estimated cost
- the market factor which could be a plus or minus quantity.

Profit is normally expressed in terms of a percentage on turnover and is therefore usually a low figure. Possibly as fair a way as any of dealing with this is to take the overall percentage profit on turnover that the contractor has achieved in his previous trading year, on the grounds that it should not be more than this when the contract is negotiated. The corresponding argument from the contractor's side would be that it should not be less because it is negotiated.

Advantages and disadvantages

Competition

The advantages of competition would be:

- accountability
- a true reflection of current market prices
- quotations which can embrace comparison of contract period, design and management.

The disadvantages of competition would be:

- a waste of resources in preparing several unsuccessful tenders
- restriction on full communication between design team and tenderers on details and site operations
- restriction on direct comparison of pricing with cost plan items during the tender period, resulting in necessary revisions of design at a later stage.

Negotiation

The advantages of negotiation would be:

- greater speed in appointing a contractor
- full communication by all parties on design, site operations and pricing of specialist items
- progressive agreement of elemental sections of the work which might enable an earlier start on site
- creation of a team spirit between professionals and contractor and elimination of many claim items during the contract period and in the final account.

The disadvantages of negotiation would be:

- a possible increase over competitive prices
- loss of time if the negotiations prove unsuccessful
- lack of accountability.

'. . . considering the many combinations . . .'

Chapter 7

Fixed Price and Cost Reimbursement

Fixed price and cost reimbursement are terms used for convenience to define the two main methods of making payment. However, their use is misleading because although they indicate the only two methods available for paying for work done, very rarely is any contract discharged entirely by one or other method. Usually a combination of both methods is employed and the contract is often named according to that which predominates.

The principles, however, remain valid. Fixed price items may be defined as items paid for on the basis of a predetermined estimate of the cost of the work, an allowance for the risk involved and the market situation in relation to the contractor's workload, the estimated price being paid by the client irrespective of the cost incurred by the builder.

Cost reimbursement items may be defined as items paid for on the basis of the actual cost of the work.

The Fixed Fee Form of Prime Cost Contract for building work offers a detailed definition of actual cost to the builder under the heading 'Definition of Prime Costs' as set down in the first schedule.

Fixed price

It will be evident from our definition of a fixed price item that fixed price can apply to a unit rate, a section of work or equally to a complete contract. Similarly, it must be appreciated that a contract may consist of a multiplicity of unit rates, a series of elemental or trade sections or a single lump sum. Understanding this should dispose of the common misconception that a fixed price contract is a lump sum contract.

This, of course, is not so and indeed a fixed price contract need not have a finite sum attached to it at the beginning of the contract. A schedule of fixed rates without quantities is a fixed price contract because the basis of payment has been predetermined – the price being fixed, only the quantity of work is unknown and this is ascertained by measurement as the work is done. This is still true even if the rate varies with the quantity done or used.

Cost reimbursement

In this system payment is not based on a predetermined contractual estimate of cost. The contractor is paid whatever the work actually costs him within the limits of the contractual arrangement, which will lay down strict rules or formulae for the calculation of that cost.

The essential difference between fixed price and cost reimbursement is that in the case of the fixed price contract the contractor contracts to do the work at a price he has estimated in advance. If he is incorrect in his estimation, then he takes the risk for being wrong. In a cost reimbursement contract, however, the building owner pays the net cost to the contractor. If that cost is higher or lower than any estimates which may have been given for the project before it was started, then the client automatically pays for the extra or gains from the savings.

Application to contract elements

We can now move on to consider in practical terms the differing elements that make up a contract and how the fixed price or cost reimbursement principles apply in each case.

Unless the contract is for a single lump sum figure to supply a specified building, it is usually made up of several elements on a fixed price contract:

- preliminaries – often a series of individual sums to represent parts of the project such as site management costs, plant, etc.
- unit rates – work fixed in place
- PC sums – nominated sub-contractors and nominated suppliers
- provisional sums – work anticipated to be required but not yet designed, including contingencies
- sums – statutory authorities' work

- profit – sometimes included in preliminaries and unit rates but now frequently separated as a sum or percentage in summary.

At once it will be seen that the first two items are of a fixed price nature while the third embodies the cost reimbursement principle. The nominated sub-contracts may in themselves be on a fixed price basis as between main contractor and sub-contractor but represent a reimbursement item as between the client and main contractor. Nominated suppliers are in the same category. Provisional sums are a way of expressing part of the client's budget in a contract and can eventually be resolved either way, and sums for work by statutory undertakings are frequently on a cost reimbursement basis. Profit, however included, is always on a fixed price basis.

Even in a 'cost plus' contract the percentage included for overheads and profit is committed in advance as being a percentage of the cost spent on the labour and materials, regardless of whether the actual cost of the overheads and management is higher or lower than this. We therefore have cost reimbursement items for labour, materials and plant but a fixed price item for management. Often a fixed lump sum fee is given for management and an estimate of cost will be made in advance as a basis for calculating this fee.

In practice, as explained above, so-called reimbursement contracts frequently contain elements of fixed price and many fixed price contracts contain elements of cost reimbursement.

Fluctuations

An element of cost reimbursement which may come into otherwise fixed price contracts concerns the payment for fluctuations in wages and materials. A distinction is frequently made here by calling a contract where fluctuations are not paid a firm price contract. When, however, fluctuation clauses are included in the contract and the contractor is paid the increase or decrease in the basic cost of labour and materials, that part of the contract is dealt with on a cost reimbursement basis where the payment is based on the market price fluctuation. However, when the payment for increases or decreases in labour and materials is based on a formula related to a national index, the basis is then fixed price and not cost reimbursement. The contractor in this case is paid on a basis fixed in advance where he takes the risk and gains if he buys in a cheaper market but may lose if he cannot.

Target cost

We have defined fixed price and cost reimbursement and indicated that they are the two basic methods of payment. However, it is possible to combine them, not only by having part of the work paid for on one basis and part on another, but by applying both principles to the same work. We have said that in a firm price contract the contractor takes the risk of his estimate being wrong, while the reverse is the case with the cost reimbursement contract. If, however, it is desired that the risk should be split between the two parties, then payment can be made in such a way as to allow this. A target cost contract is such a method and is explained in detail in Chapter 10.

Use

Just as, in discussing competitive and negotiated contracts, we suggested that negotiations should be used only where it can be shown to be more advantageous to the client than competition, so we believe that a similar principle applies here: fixed price will in the majority of circumstances benefit the client. The appropriate allocation of risk between the client and the contractor will determine the apportionment of fixed price and cost reimbursement in a contract. In assessing the situation there are two main considerations, namely the circumstances of the client and of the contractor.

The client's position

The following criteria should be considered:

- the client's financial position
- the client's time requirement for building work
- the client's corporate restrictions
- the building market.

If the client has a defined budget and adequate time for his professional advisers to obtain fixed price quotations for the work, he should pursue that route in order to minimise his risk. If the client has only £100000 to spend, his advisers should limit his risk so that this figure is not exceeded. However, if time is more important than money, the client's position may be best protected by taking more financial risk and moving towards a cost reimbursement type of contract. It is important that the client is advised on the correct balance between the two methods of payment. Any attempt to reduce the financial risk without providing adequate information to the contractor will undoubtedly backfire, with the client having to pay more as the contractor has priced for all eventualities.

The client may have a predetermined method of procurement for building work which will override his best interests. For example a Local Authority may require fixed price tenders to be obtained from an accountability point of view, yet this method may preclude the authority from having the work carried out during a particular financial year and hence may lose the money allocated for the scheme.

When work in the building industry is scarce contractors will inevitably be prepared to accept greater financial risk in order to obtain work, and the client's professional consultants should advise on that basis.

The contractor's position

The following criteria should be considered:

- the contractor's financial position
- the contractor's financially acceptable risk level
- the extent to which the work is defined
- the building market.

As with the client who has limited financial resources, a contractor in a similar situation cannot afford to take financial risks. Therefore, where he is asked to take such a risk he will add a sufficient amount to his fixed price to reduce or eliminate his risk. In certain situations he may not be prepared to take the risk at all and will refuse to tender for the work. On the other hand a contractor with substantial financial resources is in a position to take on more financial risks and submit fixed price tenders without prejudicing his immediate financial stability. Obviously he must be satisfied that the risks he is prepared to take in pricing will even themselves out over the whole range of work that he undertakes.

The financial risk level that the contractor is prepared to take is a commercial decision. Two contractors with similar financial resources may have a different tolerance level for risk. For example one may be prepared to estimate for material and labour cost increases, whereas the other will require to be reimbursed for any such increases. This action may preclude the latter from being invited to tender.

Where it is difficult or impossible to define the work clearly as in some alteration and refurbishment schemes, it will normally be beneficial for at least some of the work to be let on a cost reimbursement basis. Attempting to estimate a fixed price for work that is not properly defined will result in the items being heavily priced to cover the risk. In addition, an undefined global item will be contractually unenforceable if it proves to be priced inadequately, and conversely the client will find it difficult to reclaim any money if the work involved is less than the undefined item allows for.

Where new building techniques are being used it may be in both parties' interest to have the work carried out on a cost reimbursement basis as the true risk may not be definable. Attempting to allocate a fixed price may in these circumstances result in too low or too high a price with disastrous effects all round.

When the building market is depressed the contractor may take the view that he is prepared to take greater risks and tender on fixed prices, where in a more buoyant market he would be looking for a cost reimbursement basis.

Conclusion

When there is sufficient time it is beneficial to carry out investigative work and produce as much detailed design and specification work as possible. This work will reduce the risk element in the contract and influence the type of contract eventually used.

With the increase in use of Management Contracts it is clear that clients are prepared to take more financial risk in pursuit of shorter procurement times for building work. When a Management Contract is placed, many of the work packages will not be fixed price elements.

On all contracts, payment is made either by fixed price or cost reimbursement: FP + CR = 100 per cent. Payment can be made in no way other than under these two principles. Almost all building contracts are a combination of both.

'. . . an imaginative quantity surveyor . . .'

Chapter 8

Fixed Price Contracts

Having established general principles in earlier chapters, it is now appropriate to look at particular types of tendering procedure and contractual arrangement. In this chaper, fixed price contracts are examined.

In Chapter 7, fixed price was defined in terms of payment on the basis of a predetermined estimate, irrespective of the actual cost incurred by the builder. It was also established that a fixed price contract might consist of a single lump sum, a series of elemental or trade totals, or a multiplicity of unit rates, all with or without quantities. Fixed price contracts, therefore, appear in a number of different formats, as explained below.

Historically, fixed price contracts have accounted for the majority of building contracts and they should probably still be considered as the norm. Any departure from this norm, in favour of cost reimbursement or target cost contracts, is only recommended when the circumstances specifically require a different approach in order that the best value for money can be obtained from the client's point of view.

It is no doubt for this reason that by far the greater number of the JCT's forms of contract are essentially fixed price contracts – albeit that they usually contain optional clauses to allow an element of reimbursement of fluctuations in the prices of labour and materials. Allowing reimbursement of fluctuations passes the risk of inflation from the contractor to the employer, but other major areas of risk, such as management, supervision, resource availability and productivity, still remain the contractor's responsibility.

JCT fixed price contracts

Set out below are the JCT fixed price contracts together with the other items necessary to complete the contract documentation.

Form of contract	*Other documentation*
(1) *Standard Form (JCT 80)*	
(a) with quantities	drawings bills of quantities (incorporating the spec)
(b) with approximate quantities	drawings bills of approximate quantities
(c) without quantities	drawings specification (schedule of rates for principal items to be provided after acceptance)
(2) *Intermediate form (IFC 84)*	
(a) with quantities	drawings bills of quantities
(b) without quantities *either*	drawings schedules of work (priced)
or	drawings specification (priced)
or	drawings specification (unpriced) schedule of rates/contract sum analysis
(3) *Agreement for Minor Building Works (1980)*	
(a) without quantities *either*	drawings, with or without specification, plus schedule of work
or	spec, with or without schedule of work
or	schedules of work

There is one other form which should be mentioned, the JCT Standard Form With Contractor's Design (1981). This form is different from all the previous forms since its concept is based on a different premise, namely design provided by the contractor rather than by the client's consultants. It is the only complete form produced by the JCT which recognises design by the contractor. This different concept demands different documentation and the other contract documents in this case are the Employer's Requirements, the Contractor's Proposals and a Contract Sum Analysis.

Recognising this concept of contractor's design, the JCT also provides a Contractor's Designed Portion Supplement which is used to modify the Standard Form (JCT 80) With Quantities, to incorporate the principles identified above into the Standard Form, where part of the design is produced by consultants and part by the contractor. In this situation, the basic documentation of the Standard Form With Quantities (at 1a above) is augmented by Employer's Requirements, referring to that portion of the works designed by the contractor.

Characteristics of forms

The Standard Form (JCT 80)

The Standard Form (JCT 80) in its various versions, and with its Private and Local Authority formats, is the JCT's most comprehensive statement on contract conditions. The forms contain detailed conditions regulating the rights and obligations of the parties, the powers and duties of the architect and the quantity surveyor, and procedures appropriate to the variety of situations to be met on most projects of any size and complexity – even to the extent of covering the outbreak of hostilities and war damage, both of which make interesting reading.

The With Quantities, Approximate Quantities and Without Quantities versions, all contain similar details and differ only in respect of those conditions referring to the contract bills, remeasurement (in the case of approximate quantities) and schedule of rates (where quantities are not measured). This form, derived from a long list of predecessors, now runs to over 50 pages and is a complex document in itself. However, over the years a considerable body of case law has been built up which helps clarify the manner in which to interpret the document.

Depending on the amount of information available at tender stage, the appropriate version (With Quantities, Approximate Quantities, or Without Quantities) can be used. It should be noted that with the advent of SMM7 and Co-ordinated Project Information (CPI), the choice of form used is more directly related to the information available. No longer will bills of quantities be able to be produced without the appropriate drawings and specification information being available. In the past an imaginative quantity surveyor could produce an apparently comprehensive bill without supporting drawings and details from which the contractor could build the job, thus giving a false impression of the degree of completeness of the detailed design and construction information available.

Provision is made for nominated sub-contractors and for nominated suppliers; there is a comprehensive set of tender and contract documentation available which is fully co-ordinated with the main form of contract and which ensures that the relevant conditions are fully compatible. Although the paperwork appears complicated and many clients seek to avoid the procedure for this, and other, reasons, it only formalises the basic good practice which has always been the basis of a well run contract. The documentation should always be prepared in good time and with due care.

Provision is also made for reimbursing fluctuations in the prices of labour and materials, either by traditional means or by using the formula method when quantities are provided.

Intermediate Form (IFC 84)

This form is produced in a single format but contains a number of alternative clauses in the recitals and the appendix, which have to be selected (deleting those not required) to produce a With Quantities or Without Quantities format and to take account of a Private or Local Authority client's requirements. The form, which is set out in a clearer format than JCT 80, contains conditions that are less detailed than those of the Standard Forms but more detailed than those of the Agreement for Minor Building Works, and for this reason there is a note printed on the back of the form recommending its use for the middle range of contracts. This note suggests that the form would be suitable where the proposed works are:

- of simple content involving the normally recognised basic trades and skills of the industry
- and without any building service installations of a complex nature, or other specialist work of a similar nature
- and adequately specified, or specified and billed, as appropriate prior to the invitation of tenders.

There is also a reference to JCT Practice Note 20 (revised 1984), in which it is suggested that the form would be most suitable where the contract period is no more than 12 months and the value not in excess of £250000 at 1984 prices. However, there is a further rider noting that longer and larger contracts may be satisfactory if the three criteria noted above are met.

These limiting features could be overcome by incorporating special instructions elsewhere in the contract documentation, but in this situation it would be more appropriate to use the more comprehensive Standard Form. Amending forms of contract is never recommended where it can be avoided; apart from the sheer complexity of the exercise, case law casts doubt on the effectiveness of the result.

The provisions for the reimbursement of fluctuations in the prices of labour and materials are similar to the Standard Form, allowing for the traditional or formula methods as preferred (the formula method only being available when quantities are provided).

In the Intermediate Form an attempt is made to overcome the complexities of the nomination of sub-contractors. There are no provisions for nominating sub-contractors or suppliers, but a different and somewhat simpler concept of naming is introduced. This system allows individual sub-contractors to be selected, but once all the appropriate information is provided and a sub-contract is entered into, the specialist effectively becomes a domestic sub-contractor and the main contractor assumes total responsibility for the work (subject to various special conditions if the specialist sub-contract has to be determined). Special tender and sub-contract forms are provided to administer this process and to ensure compatibility with the main contract. It is important that this procedure is carried out in advance of the signing of the main contract, as the sub-contract has to be completed within 21 days of executing the main contract.

There is no reference in this form to Clerk of Works 'Directions' and there is no provision for excluding people from the site.

Agreement for Minor Building Works (1980)

This is again a single format document and is the simplest form available from the JCT. The form is drafted for use with drawings or specifications or a schedule of works, but no provision is made for the use of bills of quantities.

Since minor works are not considered to extend over a long

period of time, no provision is made for fluctuations in the prices of labour and materials, except for tax matters imposed by Government order.

Although specialist sub-contractors can be incorporated by naming them in the tender documents or in subsequent instructions, there are no provisions in the form for any special treatment or control of such specialists, nor are there any standard forms of sub-contract available for use in that situation.

Other, perhaps less significant omissions include:

- any provisions for insuring against damage to property other than the works
- any provision for the ownership of materials on site to pass to the employer on payment
- any provision for opening-up and testing.

As with the Intermediate Form these limitations could be overcome by drafting special instructions, but again there seems little point in doing so when there are more comprehensive forms available covering the omissions in this very simple form.

Advantages and disadvantages of fixed price contracts

The following items may be found useful in deciding to recommend to a client which of the various JCT forms of contract to use.

Advantages

- The employer avoids the risk of variations in cost due to management deficiencies, shortage of resources, reduced productivity and, to some extent, inflation, depending on which elements of the fluctuation clauses are deleted.
- Unless approximate quantities are provided the contract sum will define the client's financial commitment, subject to contingency and any other provisional allowances, and, of course, to any variations in the brief.
- There is an in-built incentive for the contractor to manage the work effectively and to complete the work as quickly as possible, so as to maximise his profit.

Disadvantages

- The premium paid to the contractor for taking the risk is paid irrespective of whether or not the risk materialises.
- Depending on the status of the contractor and the nature and extent of the risks involved, the premium required by the contractor may be excessive or poor value for money from the client's point of view.
- Time and production cost savings and improved productivity deriving from the learning curve are lost to the client; only the contractor benefits. This may be particularly significant on very repetitive jobs.

The choice between the various fixed price contracts available will depend on the other documentation available and the scope and complexity of the works. The advantages and disadvantages of the Standard Form, the Intermediate Form and the Minor Works Form are basically self-evident from their general characteristics set out above.

Operating procedures

The procedures for operating fixed price contracts have been set out in some detail in *Pre-Contract Practice* and in *Contract Administration*.

'. . . a procedure regarding checking deliveries . . .'

Chapter 9

Cost Reimbursement Contracts

In Chapter 7 cost reimbursement was defined in terms of payment on the basis of the actual cost incurred by the builder, irrespective of any estimate which may have been calculated for budgeting purposes or submitted as part of a tender. A cost reimbursement contract is therefore one in which the contractor is reimbursed his actual cost plus a fee to cover his overheads and profit.

By their very nature cost reimbursement contracts cannot have a finite sum at contract stage or when contracts are signed. The contract sum will only be ascertained at completion when the final account is settled.

Since the builder is reimbursed his actual cost it will be seen that with this type of contract the client carries all the risk: inflation, management efficiency, effective supervision, resource availability and productivity. It is therefore important that the client and his consultants have confidence in the competence of the builder and that some method of control of the builder's method of operation is included in the contract documentation.

The fee

The fee paid to the builder can be a percentage figure (applied to the prime cost) or a fixed fee (a lump sum figure). A fixed fee has the advantage of providing an incentive for the builder to work efficiently (to maximise his return). A percentage fee lacks this incentive and is sometimes considered to be an open cheque for the contractor – the more he spends, the higher his fee – and is very much akin to the concept of dayworks. Although this may be considered an extreme view, it does serve to highlight the danger of not providing an incentive for the builder to work efficiently. However, if the scope of the work is not sufficiently defined, a fixed fee may not be acceptable to the builder and a percentage may have to be agreed.

A refinement of the fixed fee method is to incorporate a provision for the fee to be varied as the final estimated prime cost varies in relation to the original estimated prime cost. This helps in dealing with variations.

JCT Contract Form

There is only one JCT standard form of contract for cost reimbursement contracts: the Fixed Fee Form of Prime Cost Contract. The current edition, incorporating revisions made in October 1976, was originally published in 1967. The comparative age of this document gives an indication that it is not perhaps one of the JCT's main stream documents. It also adds weight to the view, expressed in the previous chapter, that fixed price contracts may be considered to be the norm and that cost reimbursement contracts should only be used on those contracts where specific conditions render them more suitable, e.g. emergency repairs, investigatory works and such other cases where it is not possible to identify the scope of the work required to be done.

Characteristics of the form

The main feature distinguishing the form from that of fixed price contracts is the arrangement for payment to the contractor. Payment is based on the prime cost of the work, as defined in the documentation, and a fixed fee.

The form contains detailed conditions regulating the rights and obligations of the parties, the powers and duties of the architect and the quantity surveyor and procedures appropriate for the administration of the work. Provision is made for nominated sub-contractors and nominated suppliers but there is no detailed administrative documentation linked to this form as there is with JCT 80 fixed price contracts. Despite this there is no reason why similar up to date administrative procedures should not be incorporated independently in the sub-contract documentation.

By its very essence a cost reimbursement contract does not make any provision for reimbursing fluctuations; these are dealt with automatically, as invoices and time sheets are priced at rates current at the time the work is carried out.

Advantages and disadvantages

Since fixed price contracts are considered suitable for most circumstances there should be cogent reasons for recommending to a client the use of a cost reimbursement contract. Consideration should be given to the advantages and disadvantages of such a contract when evaluating whether or not the use would be to a client's advantage in any particular situation.

Advantages

- Work can be commenced on site immediately provided that sufficient specification information is available.
- It is often the only way of carrying out investigative work.
- It is ideal for coping with emergencies such as making buildings safe after fire damage.
- Production cost savings and improved productivity can derive from the learning curve resulting from the repetitive nature of jobs, to the benefit of the client.
- The client pays no premium to cover risk; he pays only the actual cost of whatever is deemed to be necessary.

Disadvantages

- The client's ultimate commitment is not known.
- There is little incentive to the contractor to employ his resources efficiently, but see section on the fee.
- The client carries nearly all the risk on the contract.
- Checking the prime cost can be a very complicated and expensive operation.

Budget and cost control

The procedure for creating a budget and controlling the cost in the early feasibility and design stages is similar to that for a normal fixed price contract. An early budget estimate should be prepared and there is no reason why cost planning should not go ahead normally. Furthermore, cost checking during the evolution of the design can also be carried out as with other types of contract.

It is important that the document showing the estimated cost of the project should be as detailed as possible so that the work is clearly defined and the basis on which the fee is calculated is established, whether it be a fixed fee or a percentage fee. It is necessary that the estimated prime cost should be divided between the work to be carried out by the contractor's own labour, the work to be carried out by his sub-contractors and the work to be carried out by nominated sub-contractors. This is fairly obvious, as the management expenses may be higher in respect of the contractor's own labour than of that of sub-contractors.

If it is decided that the fixed fee should be settled in competition and the contractor selected in that way, the detailed estimate must be sent to the tendering contractors as one of the tendering documents.

Administering the contract

The procedures for operating fixed price contracts (the norm) are dealt with in some detail in *Pre-Contract Practice* and *Contract Administration*. Here we consider only those procedures relating to cost reimbursement contracts which differ from the norm.

Since the standard Fixed Fee Form of Prime Cost Contract has not been updated in line with the JCT 80 fixed price contracts, it is appropriate to note that the format and content are slightly at variance with common practice and certain modifications might be thought worthwhile.

Although we do not normally advocate changing clauses in standard forms of contract, the form does contain a good deal of procedural matter which it is considered perfectly reasonable to vary in order that the most efficient method of working can be established on a particular project. For example, if joinery is likely to be supplied by the contractor's own joiner's shop, it may be desirable that the calculation of prime cost for that work should be on a different basis from the prime cost of work on site. It is also permissible to consider the incidence of small tools, perhaps even consumables, and to judge whether it is likely that fewer disputes will arise if some of these items are included in the fee rather than the prime cost.

There are other matters of procedure in which the quantity surveyor will be particularly interested. For instance, for cost control purposes he may wish to ensure that the contractor sends time sheets and invoices to him at regular intervals for checking. Often a procedure for checking deliveries and delivery notes is necessary. It is desirable to ensure that the contract covers the procedures required. These are dealt with in more detail later.

Procedure for keeping prime costs

The definition of prime cost is crucial, as generally what is not specifically included in the definition is automatically deemed to be included in the fee. With small contractors, unused to working by this method, it is wise to point this out in some detail before the contract is signed, so that there is complete understanding on both sides as to what is supposed to be included in the fee. This is the best way to avoid niggling disputes.

The quantity surveyor should agree with the contractor a working method of keeping the prime cost. Contractors are sometimes inclined to take the line that as they have been chosen to carry out work on this basis, they should have a free hand and their method of keeping the prime cost should be accepted. However, a proper system of checking the prime cost is essential and does not indicate mistrust of the contractor. It is merely a prudent method of doing business and is only right and proper on accountability grounds. Furthermore, the contractor's internal method of keeping the prime cost may not and in fact probably will not coincide with the definition of prime cost in the contract.

The quantity surveyor will, however, be well advised to go along as far as he can with the contractor's normal costing system as this is certainly the most likely to produce error-free results. There seems to be no reason why he should not receive a copy of the weekly wage-sheet, particularly where this is prepared for the contractor in any case. If some different method is used, this must be discussed with the contractor. The quantity surveyor should not object to using another method, provided that he gets the same information as the contractor with regard to wages paid.

Labour resources

It is perhaps worthwhile at the early stage, just after the contractor is selected, to agree who is going to be the working full-time foreman to be paid as part of the prime cost, and which people are in a supervisory capacity and therefore come under the management fee.

A further fundamental point arising with cost reimbursement contracts is that as the client is taking the risk, he should have some control over the way the work is carried out. For this reason it is desirable that the extent of the normal overtime to be worked is agreed in advance and that the contractor should then seek approval before working any further overtime. According to the type of project, thought should be given to all such matters and a decision made as to what can be incorporated in the documents for tendering.

Materials

A system must be worked out for the acquisition of materials. Generally speaking, competitive quotations should be obtained but it must be recognised that there may be occasions when materials are wanted so quickly that this is not possible. In such a case the quantity surveyor has to take a reasonable view and provided the materials bought are not above the market price, there is no reason why they should not be included and paid for. Even where competitive quotations can be obtained, one should consider the quality of service likely to be given to the contractor. It may be that the lowest quotation is not the most advantageous. One would, however, expect such instances to be rare.

Some system must be worked out to ensure that materials invoiced match up with materials delivered. This can be done by means of delivery tickets, and frequently contractors work such a system on their own fixed price contracts. An overall check on any particular material can be made by an assessment from the estimated prime cost. In many cases a physical check on site can easily be carried out.

Plant

The conditions of contract should state how plant is to be charged. The two main plant hire schedules in current use are:

- The RICS Schedule of Basic Plant Charges
- The Federation of Civil Engineering Contractors Daywork Schedule.

Although both these schedules are intended for use in connection with dayworks under contracts, they can, with suitable adjustment, be used for cost reimbursement contracts. Generally, it is assumed that if the plant is wanted for a long period on site it may be cheaper to buy it outright. It may be desirable therefore that some limit upon the period of hire of plant should be laid down in the contract. Sometimes this is expressed in terms of not paying more in hire than say, 80 per cent of the capital cost.

Plant hire will invariably bring with it the question of consumable stores and discussion should take place early on how these will be treated.

Credits

A system should be established for allowing credits for surplus materials and also credits for such things as old lead taken in from existing buildings. Similarly, materials supplied by the client must be taken into account when agreeing the management fee.

Sub-letting

On all cost reimbursement contracts there will be a proportion of work which is sub-let on a measured basis. This may be to a sub-contractor, nominated or otherwise, or possibly to a labour only sub-contractor. Provided the arrangement is economical and competitive there is no reason why it should not be adopted. However, if the proportion of work being sub-let rises higher than that estimated when the fee was originally fixed, there may be a case for a variation of the fee. This will have to be established before the contract is signed and, for this reason, it is desirable that the contractor's intentions regarding sub-letting on this basis should be established before he is finally selected.

Defective work

Most prime cost contracts have a clause indicating that the contractor should make good defects at his own expense. This can raise problems. If the defects are made good after practical completion there is no difficulty in separating the costs involved, but if they are made good during the progress of the work then some specific separation of the cost must be made. A kind of negative daywork charge is needed.

Cost control

Valuations for cost reimbursement contracts will normally be done on the basis of the labour paid and the invoices submitted by the contractor each month. The retention may be a percentage of the value of the work done or a proportion, say 25 per cent, of the fixed fee. There is always a considerable interval between work being carried out and invoices being submitted, and for cost control purposes, therefore, a specific reconciliation between the actual prime cost and the estimated prime cost must be made.

To do this properly, one has to go back to the detailed estimate of prime cost used for tendering and establish what proportion of that work is done. Against this, one sets the actual prime cost plus the labour and materials used but not yet invoiced. The difference between the two will show the extent to which the project is either saving on or exceeding the estimated prime cost. Obviously, in taking the original estimate of prime cost, allowance has to be made for any variations. From this reconciliation, estimates of the final cost of the project can be made. Although cost control with cost reimbursement contracts is more difficult than with fixed price contracts, it must be carried out and should be reasonably satisfactory.

Final account

The final account will, of course, be the total of the actual prime cost plus the management fee. It may be that during interim valuations various items are put into a suspense account pending settlement as to whether they are a proper charge against the prime cost or not. If possible, items in the suspense account should be settled and either included or rejected as soon as possible. The procedure for the final certificate, maintenance and handing over of drawings etc. in relation to the building is similar to that of normal fixed price contracts.

'. . . a more sophisticated calculation . . .'

Chapter 10

Target Cost Contracts

Chapter 7 concerned the relationship between fixed price and cost reimbursement contracts and established the position of the target cost contract as a half-way house between fixed price and cost reimbursement contracts. Target cost contracts are a refinement of ordinary cost reimbursement contracts and introduce an element of incentive for the contractor to operate efficiently by transferring a proportion of the risk to the contractor.

The essential feature of the target cost contract is that the difference between the actual cost and the estimated cost is split in some way between the contractor and the client. The philosophy behind this is that if the contractor cannot complete the work for the estimated cost, it is not right for the client to pay the whole of the actual cost as at least some of the increase may be due to the inefficiency of the contractor rather than to inaccurate estimating. On the other hand, if the contractor completes the work at a lower cost than that estimated, it may be assumed that some of the decrease is due to his own efficient management and, therefore, some of the gain should go to him. In this way the target cost contract gives the contractor a built-in incentive to manage the work as efficiently as possible.

It also, of course, gives him an incentive to increase the estimated price as much as possible in the first place. It is therefore essential that the client takes steps to ensure that his interests are safeguarded by employing expert advice in evaluating the estimated price. This may be negotiated with the contractor before work is started or can be established by competition as part of the original tendering process.

Target cost contracts, however, should not be entered into lightly. They are expensive to manage involving as they do both accurate measurement and careful costing on the client's behalf. This is no doubt one reason why target cost contracts are somewhat rare; however, it is important to be aware of the availability of the system should the need arise.

In a target cost contract, payment is made partly on an estimated fixed price basis and partly on the actual prime cost. This principle can best be understood by looking at two simple examples, one showing a saving on the original target and the other showing extra:

Example 1

	£
Target (i.e. estimated total prime cost including allowance for head office overheads and profit)	100 000
Actual prime cost plus overheads etc.	90 000
Therefore, saving on target	10 000
If saving is split on a 50:50 basis, payment to contractor becomes	95 000

Example 2

Target (as before)	100 000
Actual prime cost plus overheads etc.	110 000
Therefore, deficit between actual cost and target	10 000
If deficit is split on a 50:50 basis, payment to contractor becomes	105 000

There are various points to consider in studying these examples:

(1)   The figures above have been deliberately made simple; in practice the target must be revised to account for all variations. It will, therefore, be the revised target which is compared with the actual cost to establish the saving or extra.

(2)   The split can be in any proportions previously agreed. Thus if it is desired that the contractor should take more of the risk, the proportions will be agreed so that the client pays a figure nearer to the revised target and further from the actual prime cost. In the two examples given, a 25:75 split would mean that, in Example 1, the contractor would be paid £97 500, and, in Example 2, £102 500.

(3) Various methods can be used to achieve a more sophisticated calculation by splitting the work between that of the contractor and that of his own and nominated subcontractors. In addition, his head office overheads and profit can be covered by either a fixed fee or a percentage fee. In the case of a fixed fee, it would only alter if the revised target varied from the original.

(4) The target can be obtained in various ways. We have already drawn attention to the importance of reasonable accuracy here. The best way, undoubtedly, is by means of a full and accurate bill of quantities, properly priced out and agreed between contractor's and client's quantity surveyors. There may, however, be circumstances when a bill of approximate quantities will suffice. For specialist work, a method of estimating in common use by the particular trade, such as a labour and material bill, may be appropriate.

In the simple examples given, prime cost is assumed to be prime cost to the client in accordance with the detailed terms of the contract defining prime cost. For practical purposes, however, it may well be appropriate to negotiate some of the items on a fixed price basis. There are many possible variations and Example 3 (overleaf) illustrates one possible solution where three elements of the work, involving site management (5), plant (6), and overheads (7), are on a fixed price basis, while being incorporated in an overall target cost contract. Nominated subcontractors (3) and nominated suppliers (4) are included on a full cost reimbursement basis and, for the sake of simplicity, the profit or fee (8) is taken as fixed. This leaves the labour (1) and materials (2) elements (amounting to 45% of the total) subject to the target cost calculation.

Example 3

Columns A and B represent typical overspend and underspend situations

		Target	Actual	
			A	B
		£	£	£
(1)	Prime cost to the contractor of labour charges, clearly defining rates, allowances, fares, etc.	20 000	22 000	18 000
(2)	Prime cost to contractor of materials and consumable stores, making allowance for discounts	25 000	26 000	24 000
(3)	Nominated sub-contract work	15 000	16 000	14 000
(4)	Nominated suppliers	5 000	5 000	5 000
(5)	Site management (fixed)	12 000	12 000	12 000
(6)	Plant, sheds, transport, etc. (fixed)	9 000	9 000	9 000
(7)	Office overheads, supervision and insurance (fixed)	8 000	8 000	8 000
(8)	Profit or fee (fixed)	6 000	6 000	6 000
		100 000	104 000	96 000

Column A – reduction in
   profit in respect of items
   (1) and (2)
50% of £3 000   1 500

Cost to client   102 500

Column B – additional
   profit in respect of items
   (1) and (2)
   50% of £3 000   1 500

Cost to client   97 500

Competition

Very often the target will be negotiated between quantity surveyors on both sides. However, it is possible for the target to be the subject of a competition between contractors. This particularly applies in civil engineering works. In the latter case the bills of quantities will form the tendering document on which the target is based. Normally the contractor submitting the lowest target is selected.

From then on, the procedure is the same as with a negotiated target. The work is measured and valued on the basis of the bills, any variations being taken into account. The contractor is paid on the basis of the actual prime cost plus the relevant fee, and the difference between that figure and the measured account is shared between the contractor and the client on the lines indicated earlier.

Contract

There is no standard contract available for target cost work. However, the contract presents no difficulties. The standard form can be used for all but the payments clauses. It is essential to cover a definition of what is allowable in the prime cost and this, together with the procedure for keeping the prime cost, is dealt with in Chapter 9.

Advantages and disadvantages

The advantages and disadvantages of target cost contracts are very much in line with those of cost reimbursement contracts identified in Chapter 9. The only difference, and it is an important one, is that target cost contracts have the advantage of including an incentive encouraging the contractor to operate as efficiently as possible. There is a further advantage in the flexibility of the system. The risk can be apportioned in different degrees depending on the circumstances of each party and their ability to take that risk. It need never be a black and white solution; all shades of grey are attainable according to the merits of each case.

Such a contract is appropriate for high risk projects such as marine work where, although the extent of the work can be accurately defined, the conditions under which it is carried out may be beyond the control of both parties (for example, the incidence of tides and weather).

"Management contract'

Chapter 11

Management Contracts

Management contracting

Management contracting is a form of contractual arrangement whereby the contractor is paid a fee to manage the building of a project on behalf of the client. It is therefore a contract to manage, procure and supervise rather than a contract to build. Hence the managing contractor becomes a member of the client's team. In terms of the value of construction in the UK, management contracting now accounts for a significant proportion.

Payment and cost control

To understand the way a management contract works it is necessary to show how payment is made.

The building work is split into sub-contracts, generally referred to as packages. These are normally let in competition in the usual way although some may be negotiated or let on a cost reimbursement basis. All discounts usually revert to the client. It should be possible to arrange the various sub-contracts so that all building work is covered, but in practice a small site gang will sometimes be required to unload and help with the movement of materials on site and general cleaning after trades. Apart from this, it is unusual for the management contractor to carry out any of the building work himself, although he may have other companies within his group wishing to tender.

The management contractor provides the site management team and such preliminary items as offices, canteen, hoardings, etc. which are usually reimbursed at cost. On some projects, however, the cost of these forms part of his tender and is a fixed lump sum. In either case, a schedule forming part of the contract clearly defines what is required.

In addition the management contractor will be paid a fee. This fee is in respect of overheads (i.e. head office charges) and profit and is usually expressed as a percentage of the final prime cost of the works. It is therefore important to distinguish the site management team from the head office staff and facilities involved. Alternatively, the fee may be based on the cost plan and not be subject to change. Thus if the cost goes up, for no good reason, the effective percentage for overheads and profit on the actual cost decreases.

Prior to the appointment of the management contractor, a cost plan will have been prepared. As soon as he is appointed the management contractor must liaise with the quantity surveyor to prepare an estimate of the prime cost showing the estimated value of all the sub-contract packages, site management and other costs. This is then agreed by the client and architect as reflecting the required level of specification and becomes the estimate of prime cost (or EPC) for the project. Valuations and certificates for payment are prepared in the usual way. Cost reporting is carried out on a package by package basis against the EPC.

Some clients require the management contractor to be bound by a guaranteed maximum price (or GMP), this being a sum above which any expenditure will not be reimbursed. This however cannot be achieved until the design and specification of the building are fairly advanced and therefore, given the nature of a management contract, cannot usually be stated in the original contract documentation.

Selection and appointment of the contractor

Selection of the management contractor can either be on a negotiated basis or, as is more usually the case, in competition. Negotiation is easier than for a fixed price contract as the payment for most of the work relates to sub-contracts which will be dealt with separately after selection of the management contractor. Negotiation would be on the basis of the management fee together with the contractor's estimate of site management requirements.

If the contractor is to be selected in competition, tender documentation should be produced inviting several contractors to submit their proposals. This documentation will advise the contractors of everything known about the project which, given the early stage at which management contractors are often appointed, may not be very much but will generally include:

● general arrangement drawings

- known specification information
- the expected contract value
- details of any key dates
- details of the contract document.

The management contractor's submission, usually prepared in about two weeks, would include the following:

- the management fee
- an estimate (or tender if required) of the site management and preliminary costs
- any comments on the estimated cost of the work given the other information available
- a method statement giving outline proposals for carrying out the work, including a draft list of work packages (sub-contracts)
- a draft contract programme
- details of the proposed management team
- a proposed typical sub-contract form.

Clearly selection will be made largely on the credibility of the contractor to provide the building to time and to budget, as a difference of, say ½ per cent in the fee may not be significant in the context of some of the problems which could occur on a major development. With this in mind, the contractors will be anxious to include in their submissions details of their past projects.

Some or all of the contractors are generally invited to attend an interview with the client and the design team principals, after which an appointment is made. This is usually on the basis of a Letter of Intent, which may be followed up by a Pre-Construction Agreement. The purpose of the latter is to define the responsibilities of the parties and the formula for reimbursement (usually a time charge for staff and actual expenditure) in the event that the project is aborted before commencement on site.

Contract conditions

All the management contracting firms and many clients, architects and quantity surveyors have for some years now had their own forms of contract. The JCT Form of Management Contract was issued in 1987 in an attempt to rationalise the various forms in use and establish precedents. The documentation comprises the Invitation to Tender/Articles, the Conditions, the Employer –

Works Contractor Warranty (use of which is optional), Phased Completion Supplements and a series of Practice Notes.

Contract administration

In management contracting it is often the case that the whole of the management team (i.e. the design team, the quantity surveyor and the management contractor) is based or is at least represented at senior level on site. This facilitates day-to-day communication and problem solving, consistent with the objectives of fast and flexible building. Most clients will also find it necessary to designate or appoint their own project manager.

One of the first tasks is to draw up a detailed programme and procurement schedule. These show the dates by which design information is required by the quantity surveyor for the production of bills of quantities for each package. (It is usual for most packages to be sought in full competition, based on bills.) A list of possible works contractors is also compiled. These companies are then pre-qualified to ensure they are suitable, interested, financially able etc. and a number, usually four to six, are shortlisted to become the package tender list. Tenders are invited by the managing contractor and will include full preliminaries, a works-contract programme and other details such as safety and quality control procedures. Some contractors also hold mid-tender interviews to make sure the tenderers understand their part in the whole project. Any significant points arising at these meetings are, of course, circulated to all tenderers.

When the tenders are received they are very quickly reviewed by all members of the team and one, or more, of the tenderers is called to a post-tender interview. A team recommendation is then made to the client and the works contractor is appointed.

Drawings and architect's instructions are issued in the normal way. Managing contractor's instructions are issued to the works contractors within the authority of the architect's instructions and these form the basis of the package final account. Valuations are also carried out in the normal way with works contractors' applications being checked and audited by the quantity surveyor.

Periodically, reports are made by the team to the client reporting on progress, any problems being encountered or foreseen, corrective action being taken and, of course, the anticipated final account.

Professional advisors

Management contracts may include or exclude design work. The client will need independent professional advice whether the design is included or not. In the case of the architect and engineers this, of course, will vary according to where the design is placed. The quantity surveyor's role is very similar to normal building contracts but the emphasis is more on cost control and auditing of expenditure by the contractor.

Advantages and disadvantages

Traditional contracting has been described as adversarial. In management contracting, however, the sole objective of all parties is to produce the required building on time and on budget and the contractor, for his part, is paid a fee for this service in a way similar to the other members of the client's team. The process should, therefore, be more harmonious.

Management contracting allows the project to be on site earlier than by other methods. Figure 3 shows that because sub-contract

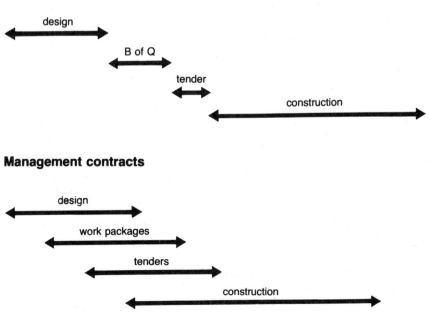

Figure 3

tenders are invited and let whilst design is still progressing, there can be more time for each of these procedures to take place whilst allowing a completion date which is earlier than by a more traditional method.

Management contracting allows the client and the designers total flexibility to develop the scheme, which enables each and every proposed design to be assessed in terms of the three factors affecting it, i.e. time, cost and quality. It is usual on a large project for the design team to be based on site.

Early involvement of the contractor means that he can advise on the suitability of proposed materials and methods with regard to time, market availability etc. Whilst the contract states a completion date, the client can, by agreement, require the works to be completed earlier although this may attract payments for acceleration of the various work packages affected. Some people express concern over cost because the contractor can be appointed whilst the design is still conceptual, but in fact almost all work packages, and many of the preliminary items, should be let following competitive tenders.

The management contractor has no conflict of interest. He is not prejudiced in any of the advice he gives, having nothing to gain or lose by the effects of that advice. There is perhaps a tendency, however, for the management contractor to wish to obtain tenders only from established sub-contractors with a reliable record. This is good for ensuring satisfactory performance but if cheapness is the most important consideration to the client, regardless of standard of workmanship and finishing times, the lowest cost may not be achieved.

Whilst the EPC is a carefully prepared and agreed budget which is constantly refined and in which confidence will grow as packages are let, it is not a commitment and some clients are unhappy about entering into a contract which does not have a contract sum.

In order to secure the management contractor's allegiance it is usual that he carries very little contractual risk, i.e. greater risk is carried by the client. For example, in the event of a default by a sub-contractor, the client would usually be responsible for all the effects of this on the cost and time of the project which could not be reclaimed from the sub-contractor.

There is a possibility of duplication of some preliminary items in cases where sub-contractors are each made responsible for an item which would otherwise have been provided by the main contractor. Scaffolding is an example.

A management contract will very likely be appropriate in the following situations:

- where it is necessary to start work on site before the design is fully developed
- where the client needs maximum flexibility to make changes to the building or to engage his own direct contractors during the course of construction
- where it may be necessary to accelerate the works.

Construction management

In many respects construction management is similar to management contracting and shares its advantages and disadvantages.

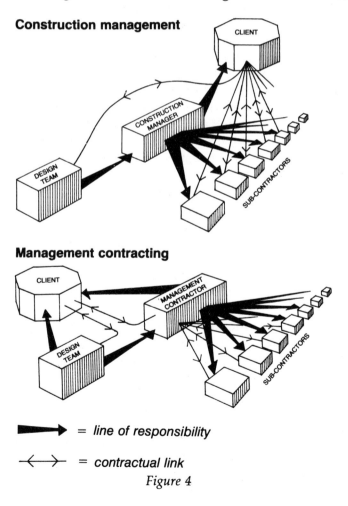

Figure 4

A construction manager is paid a fee (which may be a lump sum or a percentage of the project cost) to manage the project. None of the contracts, however, are entered into by the manager; all contracts are with the client. Whilst the manager employed will very likely be a company or partnership, it is the individual manager responsible who will affect the success of the project.

The advantage of this is that the manager, not being a party to the contracts, has virtually zero risk and therefore his sole allegiance is to the client and to the project. There is not usually a liquidated damages provision within the manager's contract of engagement.

The disadvantage is that, depending on the precise terms of engagement of the manager, the client may have no real sanction in the event of non-performance of the manager, short of replacing him. The client is a party to a large number of direct contracts and is responsible, with the manager's assistance, for the risks associated with this and the resolution of any disputes. For this reason this method of procurement is probably only appropriate for clients who regularly commission building work, have a measure of in-house expertise and wish to be involved in the detailed day-to-day progress of the works.

Figure 4 shows a diagramatic representation of the different contractual and administrative links between the parties in management contracting and construction management.

'Turnkey contracts'

Chapter 12

Design and Build Contracts

A design and build contract is a contractual arrangement whereby the contractor offers to design and build a project for a sum inclusive of both the design and construction costs.

Contractual arrangements vary considerably. They range from projects where the contractor, using his own construction staff – architects, engineers and surveyors – offers to undertake the complete design and construction, to projects where the employer appoints a design team to progress the design to a certain stage (perhaps planning approval) and then the contractor is asked to tender on the basis of completing the design and constructing the building.

Design and build contracts can be on a fixed price or cost-reimbursement basis, and can be negotiated or subject to competitive tender. They can, in certain cases, include the total financing of the project, in which case they are often described as turnkey.

Where to use design and build

There are many projects where design and build is less suitable than other forms of contract. Where architectural quality is of outstanding importance the client may rightly wish to choose the architect independently or through an architectural competition rather than use one tied to a particular contractor.

Generally, design and build does not offer many advantages in the case of refurbishment work. In large or complex projects there are many ways of bringing contractors in early to work with the design team, which lessens any special advantages that design and build might have. Clients requiring purpose-made buildings have the advantage of advice specifically related to their requirements if they commission an independent design team.

On the other hand there are standard systems of building which certain contractors may have specialised in producing, for example simple factory buildings using a proprietary system, where benefits may be obtained with design and build. Housing is an area of building where there is scope for both methods.

Where a contractor's proprietary system can be used without detriment to the client's requirements, there may well be economic advantages in a modified form of design and build. It should be noted that frequently systems of construction only apply to the basic structure; the choice of layout, finishings and external works can be to suit the client's needs without financial penalty, rather than being made to fit in with what the contractor has to offer.

Selection

The selection of a design and build contractor should be based upon a brief of the client's requirements. This brief should be carefully worked out by independent professional advisers and costed by them so that the contractors are tendering on a basis which the client knows is within his budget. It is an expensive business for design and build contractors to tender in competition as each contractor will have to work out a design to suit the brief and a price to go with it. Taken too far at the tendering stage this is clearly an example of uneconomic use of resources. In most cases contractors will not go further than an outline scheme design together with an indicative price.

Evaluation of submissions

The evaluation of the contractors' tender submissions is complex as almost certainly each contractor will interpret the brief in a different way. This means that a comparison is not being made on like with like. However, in many cases the submissions can be brought to a common denominator by valuing those items which are left out of one submission but included in others and making adjustments accordingly.

In some cases this will be a very simple and easy adjustment. For example, different floor coverings can easily be costed and if one particular contractor is eventually chosen but the floor coverings suggested by another are felt to be more suitable, it is quite simple to make an adjustment to take this into account. In other cases the evaluation will be more complicated. One design may include a pitched roof while another has a flat roof. Here it may not be possible to put the one roof on the other design, and in this case the evaluation may have to include an exercise on the ultimate cost in use of the two roofing systems.

In some cases the evaluation will be very subjective and cannot be expressed in money terms. Therefore the client's professional advisors, ideally an architect and a quantity surveyor, can only go so far in giving the client a recommendation in terms of value for money. There must always be an element of subjective advice regarding the quality of design competition.

Contract

The JCT form of contract for design and build – the JCT Standard Form with Contractor's Design – may be used. Care however should be taken to ensure that the client's professional advisors have suitable powers to monitor the work and to ensure performance of the contract.

After the selection of the contractor there will generally be a considerable period when the contractor's designers are refining the design and taking into account the particular requirements of the client where these have differed from the brief or have been misinterpreted. Frequently the evolution of the design will bring to light matters which at that stage can be incorporated in the building without penalty and to the client's benefit. All this must be taken into account when finalising the specification and performance requirements of the building and the final contract price. This is a matter for negotiation between the client's professional advisers and the contractor's design team.

Post-contract administration

The responsibility for carrying out the design will rest with the contractor's management team and his own design staff. The client's professional team will be concerned with monitoring the work and ensuring that the specification is adhered to as set out in the contract documents and that the performance is up to the agreed standard.

Care must be taken to ensure that the contractor is paid the correct amount in the way of interim and final payments. In this respect it is essential that the client appoints his own quantity surveyor to act directly on his behalf. Valuations for payments on account will be made by the client's quantity surveyor in the usual way and of course any variations requested by the client will have to be valued by the client's quantity surveyor and negotiated with the contractor so that they can be incorporated in the final account. To facilitate this, the contractor should be required to provide cash flows and a breakdown of his tender figure.

Advantages

- An early certainty of overall contract price is obtained.
- Responsibility for the performance of the building lies entirely with one party – the contractor.
- Design and build imposes a discipline, not least on the client, to define the brief fully at an early stage. Given this, the advantages of overlapping design with construction can lead to a shorter project duration.

Disadvantages

- Design and build is by its very nature a rigid system, which does not allow the client the benefit of developing his requirements and ideas.
- Where several tenders are invited, comparison of these can be difficult as the end production in each case will be different and subjective judgements will apply.
- Any variations required by the client after signing a contract can prove expensive and difficult to evaluate.
- There is always a risk with regard to the quality of the work. If the original brief is not precise and the specification offered by the contractor equally vague there is a temptation for the contractor to reduce standards.
- Most design and build contracts are qualified in some respect (e.g. ground conditions, the inclusion of provisional sums), which to a certain extent negates the client's ideal of a known commitment.

'. . . small errors of documentation . . .'

Chapter 13

Continuity Contracts

Various procurement and contractual methods available when dealing with one project or construction site have been discussed in previous chapters. In circumstances where continuity is required, advantage can be gained by adopting a different approach. This can also be the case on maintenance contracts which require the same repetitive type of work to be undertaken on a number of buildings and/or for a certain period of time. In this chapter we compare the purpose and use of three types of continuity arrangements and describe their operation to illustrate how benefits can be gained. These are:

- serial contracting
- continuation contracts
- term contracts.

Serial contracting

In many programmes of building work, such as providing several primary schools for an education authority or refurbishing a high street shopping chain for a retailer, there is an element of continuity. The object of any tendering procedure should be to ensure that building resources, including associated professional resources, are used as economically as possible. Where there is continuity these resources may be used more economically and efficiently if all the work is carried out by the same contractor rather than by having a different one for each site or operation. It is important, therefore, to identify where the continuity lies and to evaluate the benefits that will accrue if one contractor undertakes the work instead of several.

However, it is not enough merely to establish that there is a saving in resources. As the client has provided the continuity by virtue of his programme, he should gain financially in addition to securing consistent quality and good working relationships. It will be appreciated that in any situation where one contractor is to do a series of jobs for a client, a good relationship beomes particularly important.

Serial contracting has been broadly defined as an arrangement whereby a series of contracts is let to a single contractor, but further identification is needed to distinguish it from other types of continuity arrangement. In a serial contract the approximate extent of the series is known when the offers are obtained, and the individual projects tend to be of the same order of size. The serial tender is a standing offer to carry out a series of projects on the basis of pricing information contained in the competitive tendering documents. These may consist of bills of quantities specifically designed as master bills to cover the likely items in the particular projects envisaged.

The series will usually consist of a minimum of three to a maximum of perhaps fifteen projects. The number will normally be known at the time of the tender, although further projects may be added by agreement at a later stage. Final designs will not necessarily be ready for all of them at tendering stage but the client and his architect will have a good idea of the typical requirements, and these will be reflected in the tendering documents prepared by the quantity surveyor.

Exact quantification will be achieved as the design for each project in the series is finalised. Normally a separate contract, based on the original standing offer, will then be negotiated and agreed for each project based on standard items extracted from the master bills and all new or 'rogue' items being negotiated for each contract as they occur.

Purpose and use

Serial contracts are ideally suited to a programme of work where the approximate number and size of projects to be contracted is known at the time of going to tender. For example, in refurbishing a series of high street shops for one client it will normally be the quantity of work which varies rather than the quality or specification. Similarly, although primary schools for an education authority may vary in size, in a new building programme the largest is likely to be less than three times the size of the smallest. Moreover, the sizes of most of the projects are likely to be far closer to each other than this. Therefore the same type of contractors' management is likely to be needed for them all. It is probably this latter point which should be regarded as one of the governing criteria.

Consideration should be given to the choice of this method of contracting in times of high inflation, or when there is a building boom, when it may be difficult to secure a firm price commitment over a lengthy period of time.

Operation

On serial contracts the tendering procedure will normally be on the basis of a master bill, which will comprise most of the items envisaged in the various projects in the series. Individual projects will be controlled by having separate contracts negotiated on the basis of the master bill, after which work will proceed very much as for a normal contract.

It is desirable to take great care in the selection of the contractors to tender before formal tendering takes place. A mistake made when dealing with a series is likely to have more serious results than in the case of one-off contracts, so it is much more important to establish the financial and physical resources of the contractor. A bankruptcy, covering as it would many contracts rather than one, would be disastrous. It could be advantageous to make provision within the contract for the client to be able to withdraw from a contract if performance is found to be below par. A further point to note is that small errors of documentation or detail which may not amount to much on one contract are clearly multiplied in a serial situation.

Comparing the contractor's capacity with the likely programme of work also requires great care. The individual contracts may be relatively small although the total series may be large. It is possible therefore that a contractor who might not be able to cope with such a large contract on one site would be able to cope with a number of smaller projects spread over several sites and, possibly, over a longer time. It is important however that his physical resources should not be stretched to the point where it is difficult for him to complete. Any unprogrammed overlaps in projects could present a contractor with resourcing problems. If significant changes in the overall programme occur it may be worth considering retendering.

It is likely that, before the formal tendering, there will be a stage where interviews are held with possible tenderers, references sought, and a short-list drawn up on the basis of the information obtained. The result of competition should be that the saving in building resources arising from the continuity is passed to the client. It is emphasised, however, that this is only possible where the extent of the work to be covered by the serial contract is known sufficiently for reasonable estimates of prices to be given in the original standing offer.

Continuation contracts

A continuation contract differs from a serial contract in that it results from an *ad hoc* arrangement to take advantage of an existing situation. Thus, if an office contract is already going ahead on a particular phase of a business park site, the opportunity may arise to commence a further phase with similar requirements. In this case there may have been no standing offer to do more work, the original tendering documents having been conceived only for one particular phase. However, they can provide a good basis for a continuation contract. Alternatively, it is possible to make provision for continuation contracts in the tendering documents for the original project, and this is sometimes done. There is, however, no contractual commitment and the possibility of a continuation contract may not arise.

In either case the contractor has to price the tender documents for the original contract on the basis of getting that particular contract only. The continuation contract, if and when it arises, is then dealt with separately.

Purpose and use

As with serial contracts, continuation contracts are best suited to situations where the scope, style and nature of the work are very similar. The more variations there are between the projects, the more variations there will be in the contract documentation.This will result in an increased number of purely negotiated costs (rather than adjustments of competitive prices) and could also lead to organisational problems on site.

The purpose of a continuation contract would be to seek a financial advantage for the client. Continuation contracts offer the client the following potential benefits:

- a faster start on site, resulting from the shorter pre-contract period

- a competitive basis for pricing, resulting from the initial tender
- an experienced contractor, having established the construction problems of the previous contract (this represents perhaps both a cost and a time benefit)
- re-use of the contractor's site organisation and his management team

Whilst circumstances do arise where the use of continuation contracts might be appropriate, such as phases of a business park, the practicalities of a situation often preclude this. For example, whilst the style of the work is very similar the projects would not necessarily be of a comparable size. Additionally uncertainties such as planning approvals, existing leases/ tenancies and enabling/preparatory works often make site acquisition and site handover ready for construction a precarious job and create obstacles to continuity.

Provided there is a degree of parity between the projects however, a continuation contract may be appropriate where serial contracting is not.

Operation

In continuation contracts the tendering documents will obviously be closely related to the documentation for the original project on which the continuation is to be based. The contract sum is negotiated on the basis of the original contract but with two substantial adjustments.

Firstly, the contractor may require increases in costs of labour and materials to be taken into account to allow for the difference in time between the two tenders. The only exception to this might be where the original contract sum was on a fluctuations basis; the continuation contract could then, if necessary, be from that same base, in which case the increases would be covered entirely by the fluctuations clause. The same would apply, of course, to decreases if any.

Secondly, the client will wish the benefits of the continuity he has provided to be passed to him in the continuation contract. These will be of two kinds:

- Productivity in the construction industry generally has increased year by year and, therefore, if the continuation contract is timed say, 9 months to a year later that the original contract, there should be a saving resulting from increased productivity.

- In additon to this, however, there is a further saving arising from the economies inherent in the same contractor doing similar work for the same architect and client. This element varies from one contract to another but may be quite substantial where similar houses or flats are being built on the second site.

The matter of measurement of productivity is difficult and requires much research. A lot will depend on the information available to the quantity surveyor, his skill in analysing and assessing elements which affect productivity, and in negotiating afterwards.

One of the prerequisites for the successful negotiation of a continuation contract is parity of information between client and contractor. This involves disclosing the details of how the tender for the original contract was built up. If the contractor is not prepared to do this then it may be prudent not to proceed on this basis.

Once the continuation contract has been negotiated, a figure agreed and a contract signed, the project will proceed as any other. There is one factor, however, that should be considered, particularly where it is envisaged that several continuation contracts may arise from the original. The contractor should be given an incentive to make cost reductions. Where continuation contracts are being considered the contractor might well be given the full benefit of any reduction in cost he makes, even though this involves a change in specification (but not, of course, in performance), with the proviso that the full amount of the cost reduction be allowed to the client in the succeeding continuation contract. This situation arises particularly where the contractor is involved in a considerable amount of design work and where the detailed specification is very much related to his production requirements.

Term contracts

Term contracts differ from both serial and continuation contracts in that they envisage a contractor doing certain work for a period of time or term. In this situation the contractor agrees a contract to do all work that he is asked to do within a certain framework and during a given period. Term contracts are generally set to run for a period of twelve months but a longer period is likely to result in better tenders. Experience suggests that a term of up to three years can generally be expected to promote confidence within the period of the contract. However, the volatile nature of building costs is such that provision should be made for the employer to test the market at shorter intervals. Thus a period of two years may, on balance, be regarded as the optimum for such contracts. Provision can be made for revising unit rates regularly, relative say, to the annual rate of inflation.

Purpose and use

Term contracts are well suited for use with the management of large and continuing programmes of day-to-day reactive building maintenance. The management and control of such programmes can be an extremely complex task and in view of their potential for flexibility, term contracts are perhaps best suited to the problem.

Care should be taken, both in the compilation of tendering documents and in their evaluation. Generally speaking the tendering document will be a schedule of rates. Additionally, it can be helpful to prepare a bill of provisional quantities, based on previous years' workflow, to give an indication of the likely volume of work anticipated on specific trades over the term of the contract. As always it is advantageous if contractors of like characteristics and performance can be selected to submit tenders and it will usually be an advantage if multi-trades contractors can be utilised in view of the many operational, as opposed to single trade, jobs to be found in building maintenance.

Term contracts may also be used for painting and redecoration, road-works and other specialist trades.

JCT Contract Form

At the request of its constituent bodies, the JCT produced the Standard Form of Measured Term Contract in 1989. It is specifically intended for clients who have programmes of regular maintenance and minor works including improvements. The JCT have also issued a Practice Note and Guide relating to the form.

This form requires the client to list the properties in the contract area and the type of work to be covered by the contract, as well as the term for which it is to run. The client is required to estimate the total value of the contract and to state the maximum and minimum value of any one order. There is provision for priority coding of orders to deal with emergencies and specific programming requirements.

Payment is made on the basis of the National Schedule of Rates, or some alternative priced schedule, and the contractor quotes a single percentage adjustment to the base document. Provisions are included for fluctuations and dayworks. Measurement and valuation of individual orders can be carried out by the contract administrator or by the contractor, or can be allocated to either party according to value.

Operation

Term contracts are ideal for carrying out maintenance and repair work. In this type of work the individual project can be very small, say from £5, while the largest may be in the region of £30 000 or more. Tenders can be sought using a large number of methods. Two of the commonest are as follows.

The client can determine the unit or operational rate for each item and the tenderers then offer to do the work on a plus, minus or zero percentage basis to reflect their bid. Alternatively tenders can be sought on the basis of blank schedules which the tenderers then price. Analysis of the tenders will usually be found to be easier if the first method is used; depending on the number of tenders received the analysis of priced schedules can be a daunting task. Experience will suggest the best method for individual circumstances. For example, a pre-priced schedule which includes a number of blank items for multi-trade jobs, to be priced separately by the tenderer, can result in better value for money for the client where the price for the job is less than the sum of the component operations. Additionally, as part of their contract, tenderers may be required to meet specified response times for various categories of work, including emergency items. The ability to meet these response times will be an important consideration in the selection of the successful contractor(s).

Contracts can be entered into with one or more contractors and the work placed on the basis of cost, committed workload and performance. Generally speaking the more contractors available to do the work the smoother the operation of the maintenance programme.

Orders for work are issued from time to time, the work is agreed and the contractor paid accordingly. The number of orders issued annually under a maintenance term contract can run into many thousands. To ensure the work loads can be dealt with smoothly and expeditiously continual monitoring of the contractors' performance will be necessary and orders to the various contractors regulated accordingly.

Orders should be given a priority rating which will require contractors to attend and carry out the work within a specified time. An incentive scheme will be found to be helpful to achieve this. Where the work is completed within a specified time an attendance payment can be made. Alternatively, and perhaps more effectively, where the contractor fails to complete the job on time the number of orders can be reduced until performance improves again. Additionally a plus payment can be made for attendance to emergency call-outs. Emergencies, which relate to health and safety matters, can arise at any time and will usually require quick attendance, initially to make safe only; the permanent repair can be dealt with separately.

The average cost of day-to-day maintenance items will be found to be something up to £100. Assuming that the builder deals, on average, with say, 100 orders per week, his average monthly account will be in the order of £35 000 to £40 000, a not inconsiderable cashflow. It is therefore incumbent on the employer to ensure that payment of the builders' accounts is made expeditiously.

Receiving repairs requests, deciding on remedial action, determining priorities, allocating work, monitoring attendance and completion times and inspecting and certifying payment can be a complex exercise. The use of a sophisticated computer programme based on a property register, linked with a comprehensive schedule of rates, an order generator and accounts for payment will be found to be invaluable in the smooth running of a maintenance based term contract.

For a client with a large estate and a large number of minor works of this nature to be carried out, a term contract is a method of controlling the work with a measure of accountability and a simple method of procurement for the individual project.

'. . . incentive . . .'

Appendix A

Glossary of Terms

Many terms used in tendering and contract are words in common use in the English language which have been given a particular meaning in this context. The definitions below are deliberately kept brief, as in many cases a full discussion of the meaning is to be found in the text. It is hoped that these brief definitions will help the reader when he encounters the word before it has been fully discussed.

ACCOUNTABILITY
The responsibility to justify, to those on whose behalf contracts are let, the procedures and decisions for the selection and payment of contractors and for the administration of the contract.

BILLS OF APPROXIMATE QUANTITIES
As bills of quantities, but the quantities are approximate – note that the descriptions of work are still accurate.

BILLS OF QUANTITIES
The quantified measurement and description of work in a proposed building contract – a contract document in the Standard Form.

BUILDING RESOURCES
The physical resources needed to build in terms of labour, materials, plant, management, consumable stores and finance.

COMPETITION
The endeavour, simultaneously with others, to gain a contract – applicable to both main and sub-contracts and used in respect of both selection of the contractor and determination of price.

CONTINUITY
The repetition of similar types of work or management of work; continuation contracts arise from this situation but see also serial tendering for distinction.

CONTRACT DOCUMENTS
The documents signed by both parties and precisely stated in the form of contract as a 'Contract Document'. Contracts can be signed under hand or sealed.

CONTRACT PERIOD   The length of building time on site between the date for possession of the site to the date of completion stated in the contract or the certificate of practical completion.

CONTRACT SUM   The sum as defined in the contract to be paid to the contractor for work to be carried out.

CONTRACTUAL ARRANGEMENTS   All those arrangements in the contract documents showing the obligations, rights and liabilities of the parties.

COST   Strictly the cost to the contractor as distinct from the price to the client. However, sometimes used interchangeably with price.

COST CONTROL   In spite of the definition of cost, the term normally refers to the control of cost to the client (i.e. price) from budget right through design and construction to final payment.

COST PLAN   A plan formulated by the quantity surveyor in which the estimated cost of a building is apportioned among the various construction elements to enable control of the cost to be exercised and to allow, within each element, comparison of various methods of construction and specification.

COST PLUS   A contractual arrangement whereby the contractor is paid the net prime cost plus a percentage to cover overheads and profit.

COST REIMBURSEMENT   Reimbursement by the client of the net costs the contractor has paid out – one of the two fundamental methods of payment to contractors, the other being fixed price. In practice, nearly always modified to include an element of fixed price, i.e. obtaining competitive quotations for specialist sub-contract work.

DEMOLITION CONTRACT   A contract for the demolition of a building or other structure let separately from the main building contract and normally carried out by a specialist demolition contractor.

DESIGN AND BUILD CONTRACT   Design/build contracts are tender and contractual arrangements whereby the contractor offers to design and build a project for a sum inclusive of both the design fee and construction costs.

DETERMINATION OF PRICE	The establishment of the price to be paid for work – often bound up with selection of the contractor but not necessarily so.
ELEMENTAL BILLS	Bills of quantities presented in elemental form rather than trade by trade. The elements normally represent the function of the construction as far as practicable and will be used for cost analysis.
FIRM PRICE	A contractual arrangement where labour and material price fluctuations are at the entire risk or benefit of the contractor. However, changes in government taxes are given or paid in a firm price contract.
FIXED FEE	A fee paid generally for management, fixed in advance where the remainder of the contract is on a cost reimbursement basis; such a contract is sometimes referred to as 'cost plus fixed fee'.
FIXED PRICE	One of the two fundamental methods of contract payment (the other being cost reimbursement), based on estimated figures established in advance of the work being carried out, which includes all the allowances required by the contractor to undertake the work including risk, factors reflecting the state of the market and the contractor's workload. May be in respect of the total work, a part, or units of work.
FLUCTUATIONS	The increases or decreases in cost of employing labour and purchasing materials – not variations which relate to changes in work in a contract.
INCENTIVE	Motivation to achieve – used particularly in contracts in respect of the possibility of the contractor making reductions in the cost of carrying out work.
LUMP SUM	A fixed price, but in a single sum for the total contract work and not intended to be adjusted by variation or remeasurement.
MANAGEMENT CONTRACTING	Management contracting is a form of contractual arrangement whereby a contractor is paid a fee to manage the building of a project on behalf of the client. It is, in essence, a contract to manage rather than a contract to build.

NEGOTIATION	The settlement of prices by agreement. Negotiated contract: a contract price based on negotiated prices instead of being awarded as a result of competition. Negotiation usually refers only to the main contractor's own work as sub-contract work is often awarded through competition.
NOMINATED SUB-CONTRACT	A sub-contract where the sub-contractor is nominated by the architect or engineer and not by the contractor.
OPERATIONAL BILLS	Bills of quantities presented in a form reflecting the likely operations on site for each section of work.
PARALLEL WORKING	A method of working whereby the production drawings will be produced in parallel with work on site in accordance with an agreed programme.
PARITY OF INFORMATION	The free exchange between parties of estimating information which might be considered confidential in a competitive situation, but is often a condition precedent to negotiation.
PERFORMANCE SPECIFICATION	A specification of work related to the performance required (e.g. operation, capacity, strength, etc.) rather than in terms of the workmanship and materials to be used.
PRICE	The cost to the client – normally in a fixed price contract will be based on the contractor's estimate of his own cost plus what he thinks he can obtain as profit in relation to the current market situation.
PRIME COST	The net cost to the contractor of a particular item or piece of work. The initials PC used in bills of quantities and specifications have a particular contractual implication denoting work to be carried out by nominated sub-contractors or suppliers and statutory authorities.
PRODUCTION COST SAVINGS	Savings in the contractor's costs of carrying out the work – productivity savings refer to methods of carrying out work more efficiently, particularly in respect of the use of labour and plant.
PROPRIETARY SYSTEM	A system of building designed by the contractor or sub-contractor and usually used by him alone or by others under licence.

PROVISIONAL SUM	A sum included in a tender and contract for work anticipated but not sufficiently designed or detailed to permit definitive description and measurement. The description of the provisional sum should, however, define the item of work sufficiently to enable the contractor to allow accordingly in his programme within the contract period and to understand the basis for detailed price assessment in the final account.
RISK	The hazard inherent in forecasting the cost of work to be carried out at a future date which results in exposure to loss. Either party to a contract can agree beforehand to accept the risk of such loss for the whole or part of the work. If the contractor accepts it he will normally add something to allow for it. If the client accepts it he pays the actual cost, not the estimated cost.
SCHEDULE OF RATES	A list of unit items of work priced at a rate per unit. The schedule thus formed is used in conjunction with the measurement of work to calculate payment.
SELECTION OF CONTRACTOR	The choice of contractor; while often bound up with the determination of price, it is not necessarily so.
SERIAL TENDERING	A standing offer to carry out work for more than one project in accordance with the tender submitted for the initial project, or based on hypothetical bills of quantities representing the average project of a series. Serial contract is the contract relating to such a tender.
SITE MANAGEMENT	The personnel carrying out management work full-time on site as against head office staff who will be involved only part-time and deal with activities for the firm as a whole.
SPECIFICATION	Unquantified description of the work required to be done. For building work under the JCT Standard Form of Contract With Quantities, the specification is incorporated in the bills of quantities and is not a separate contract document.
STANDARD FORM OF CONTRACT	The form of contract for building work published by the Joint Contracts Tribunal, referred to as the JCT Form. The constituent bodies of the JCT and the types of forms available are listed in Appendix B.

SUB-CONTRACTOR (DOMESTIC)
A sub-contractor selected by the contractor, sometimes from a short list of sub-contractors listed by the architect in the tender document.

SUB-CONTRACTOR (NOMINATED)
One specified sub-contractor selected by the architect in accordance with set rules laid down in the contract conditions.

SUBSTITUTION BILLS
In contracts based on bills of approximate quantities, bills prepared as production information becomes available and substituted for the original approximate bills.

TARGET COST CONTRACT
A contract in which the contract sum is calculated by comparing the actual prime cost with an estimate of the cost (target) made in advance, the difference between the two being shared between the parties according to the actual terms of the contract.

TENDER
An offer to carry out work at a price – a bid in accordance with the conditions set down in the tender documents.

TENDER DOCUMENTS
All documents upon which the tender is based. Quite often the tender documents are incorporated in the contract as 'contract documents'.

TENDERING PROCEDURE
The process whereby a contractor is selected to carry out work and the basis of settlement of an offer on which a contract may be let.

TERM CONTRACT
A method of carrying out work for a fixed term on the basis of a schedule of rates – usually used for alteration and repair work by large organisations and in the public sector.

TURNKEY CONTRACT
A contract where the contractor provides the total resources required including the finance as well as design, construction and fitting out.

TWO-STAGE TENDERING
Commonly used to indicate situations where a competition takes place to select a contractor at an early stage on a basis which, because of the lack of information then available, precludes the determination of the contract price, this being settled later at the second stage.

VARIATIONS
The contract term for changes in work authorised by the architect on behalf of the client – to be distinguished from fluctuations.

Appendix B

Standard Forms of Building Contract

The most common forms in use for building work are those issued by the Joint Contracts Tribunal, whose constituent bodies are:

Royal Institute of British Architects
Building Employers Confederation
Royal Institution of Chartered Surveyors
Association of County Councils
Association of Metropolitan Authorities
Association of District Councils
Confederation of Associations of Specialist Engineering Contractors
Federation of Associations of Specialists and Sub-Contractors
Association of Consulting Engineers
British Property Federation
Scottish Building Contract Committee

The Joint Contracts Tribunal have issued and regularly amend the following Standard Forms of Contract with supporting documentation and Practice Notes. The standard forms issued by the Joint Contracts Tribunal in 1963 should no longer be used.

JCT Standard Fixed Fee Form of Prime Cost Contract (1967)

JCT Standard Forms of Building Contract (1980)

Local Authorities with Quantities
Local Authorities without Quantities
Local Authorities with Approximate Quantities
Private with Quantities
Private without Quantities
Private with Approximate Quantities
Fluctuation Clauses
Formula Rules
Sectional Completion Supplement
Contractor's Designed Portion Supplement (1981)
Agreement for Minor Building Works
Nominated Sub-Contract Tender and Agreement – NSC/1
Employer/Nominated Sub-Contractor Agreements – NSC/2 and 2a
Form of Nomination of Sub-Contractor – NSC/3

Nominated Sub-Contracts – NSC/4 and 4a
Nominated Supplier Tender – TNS/1
Nominated Supplier Warranty – TNS/2
Domestic Sub-Contract Articles of Agreement – DOM/1
Domestic Sub-Contract Conditions – DOM/1

JCT Standard Form of Building Contract with Contractor's Design (1981)
Standard Form of Contract – WCD
Formula Rules
Domestic Sub-Contract Articles of Agreement – DOM/2

JCT Intermediate Form of Building Contract (1984)

Standard Form of Contract – IFC
Fluctuation Clauses
Formula Rules
Sectional Completion Supplement
Form of Tender and Agreement for Named Sub-Contractor – NAM/T
Named Sub-Contract Conditions – NAM/SC
RIBA/CASEC Form of Employer/Specialist Agreement – ESA/1
Domestic Sub-Contract Articles of Agreement – IN/SC (1985)
Domestic Sub-Contract Conditions – IN/SC (1985)

JCT Standard Form of Management Contract (1987)

Standard Form of Contract
Works Contract/1 – Section 1 – Invitation to Tender
Works Contract/1 – Section 2 – Tender by the Works Contractor
Works Contract/1 – Section 3 – Articles of Agreement
Works Contract/2 – Works Contract Conditions
Works Contract/3 – Employer/Works Contractor Agreement
Phased Completion Supplements
Formula Rules

JCT Standard Form of Measured Term Contract (1989)

The following forms have been published by the Association of Consulting Architects and the British Property Federation:

ACA Form of Building Agreement (1982)
ACA Form of Sub-Contract (1982)
ACA Form of Building Agreement BPF Edition (1984)

The following forms are used for Government contracts:

General Conditions of Contract for Building and Civil Engineering – GC/Works/1 Edition 3 (1989)
General Conditions of Government Contracts for Building and Civil Engineering Minor Works – GC/Works/2 Edition 2 (1980)
Sub-Contract Articles of Agreement and Appendix – GW/S (1985)
Sub-Contract Conditions – GW/S (1985)
Priced Adjustment Formula Provision – GW/S (1985)

There are also Standard Forms of Contract which are published specifically for civil engineering works. Conditions of Contract and Forms of Tender, Agreement and Bond for Works of Civil Engineering Construction (commonly known as ICE Conditions of Contract) issued by:

Institution of Civil Engineers
Federation of Civil Engineering Contractors
Association of Consulting Engineers

Conditions of Contract (International) for Works of Civil Engineering Construction approved by:

Fédération Internationale des Ingénieurs-Conseils
Fédération Internationale des Entrepreneurs Européens de Bâtiment et de Travaux Publics
International Federation of Asian and Western Pacific Contractors' Associations
La Federación Interamericana de la Industria de la Construccion
The Associated General Contractors of America

Other so called standard forms of contract have been issued unilaterally by specialist organisations. If used at all, great care must be exercised as the client's interests may not be protected. It is also worth noting when working overseas that some countries have their own standard forms of contract which may need to be considered, although many of these were originally based on the contracts listed above.

There follows a reproduction of the *Simple Guide and Flow Chart for Selecting the Appropriate JCT Form of Contract,* published by the BEC, which may be found useful in reviewing the choice of forms available and selecting the one most suited to any particular set of circumstances.

SIMPLE GUIDE AND FLOW CHART* FOR SELECTING THE APPROPRIATE JCT FORM OF CONTRACT

*(See reverse for Flow Chart)

Designer of the Works	Contract Documents	Value Range	Recommended Form of Contract		Comments	Type of Contract	Fluctuations available	Subcontract forms		Comments
			Private Sector	Public Sector				Nominated	Domestic	
Architect or other Professional engaged by Employer	Drawings and full Bill of Quantities	£250,000 (1980 Prices)	JCT 80 Private with Quantities (80/PW)	JCT 80 Local Authorities with Quantities (80/LAW)	For general use and where works are of a complex nature or with a high content of building services or other specialist work or exceed 12 months duration.	Lump Sum	Tax: Full: Formula.	NSC/4 or NSC/4a	DOM/1	NSC/4 used where S/C has tendered on NSC/1, executed NSC/2 and been nominated on NSC/3 under the Basic Method. NSC/4a used where S/C is Nominated under the Alternative Method
Ditto	Drawings and Specifications	Up to £100,000 (1980 Prices)	JCT 80 Private without Quantities (80/PWQ)	JCT 80 Local Authorities without Quantities (80/LAWQ)	Not recommended if works are of a complex nature that can best be explained to the tendering Contractor by use of a Bill of Quantities.	Lump Sum	Tax: Full: Formula.	NSC/4 or NSC/4a	DOM/1	Ditto
Ditto	Drawings and Approximate Bill of Quantities	General	JCT 80 Private with approx. Quantities (80/PWA)	JCT 80 Local Authorities with approx. Quantities (80/LAA)	For use where extent of work is not fully known (eg alterations or repairs) or where an early start is required before detailed contract documents can be prepared.	Re-measure	Full: Formula.	NSC/4 or NSC/4a	DOM/1	Ditto
Ditto	Drawings and Specifications or Bill of Quantities	up to £250,000 (1984 Prices)	JCT Intermediate Form of Building Contract (IFC/84)		For simple contracts not exceeding 12 months duration without high content of building services or other works of a specialist nature.	Lump Sum	Tax: Formula	N/A	NAM/SC or IN/SC	NAM/SC is compulsory where S/C has been named by Architect IN/SC is for use where S/C appointed by Contractor
Ditto	Drawings and Specifications or Schedules	up to £50,000 (1981 Prices)	JCT Agreement for Minor Building Works (80/MW)		For simple contracts of short duration.	Lump Sum	Tax only	N/A	N/A	
Ditto	Specifications with or without Drawings	General	JCT Fixed Fee Form of Prime Cost Contract		For use where nature or extent of works cannot be fully described or where an early start is required before design is finalised.	Actual Cost plus Fixed lump sum Fee	N/A			
Contractor	Employers Requirements. Contractors Proposals. Contract Sum Analysis	General	JCT Standard Form with Con-Contractors Design (81/CD)			Lump Sum		N/A	DOM/2	
Part Architect or other Professional engaged by Employer and part Contractor	Drawings and Bill of Quantities plus Employers Requirements for the Contractor Designed Portion.		JCT Private with Quantities (80/PW) plus Contractors Designed Portion Supplement (81/CD/DP)	JCT 80 Local Authorities with Quantities (80/LAW) plus Contractors Design Portion Contractor (81/CD/DP)	For use with JCT-80 Standard Forms where part of the works (eg Piling, suspended floors, roof trusses, etc) are to be designed by the Contractor	Lump Sum	Tax: Full: Formula	NSC/4 or NSC/4a	DOM/1 or DOM/2	

Produced by the BEC 82 New Cavendish Street. London W1M 8AD. Telephone 01-580 5588. Telex 265763

Reproduced by permission of the Building Employers Confederation.

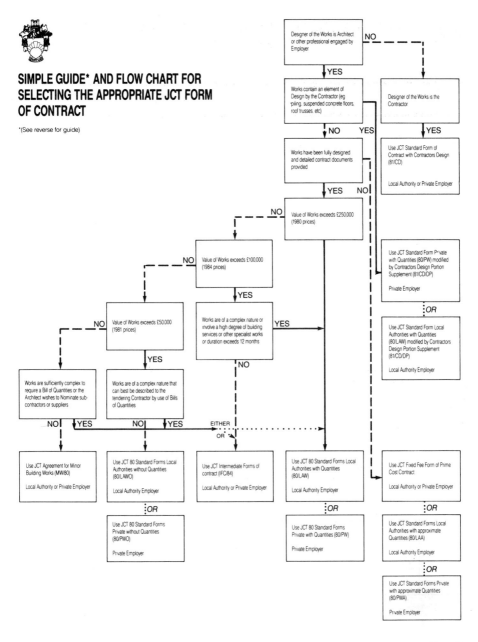

SIMPLE GUIDE* AND FLOW CHART FOR SELECTING THE APPROPRIATE JCT FORM OF CONTRACT

*(See reverse for guide)

Designer of the Works is Architect or other professional engaged by Employer — NO / YES

Works contain an element of Design by the Contractor (eg piling, suspended concrete floors, roof trusses, etc) — NO / YES

Designer of the Works is the Contractor — YES

Use JCT Standard Form of Contract with Contractors Design (81/CD)

Local Authority or Private Employer

Works have been fully designed and detailed contract documents provided — YES / NO

Value of Works exceeds £250,000 (1980 prices) — NO

Value of Works exceeds £100,000 (1984 prices) — NO / YES

Value of Works exceeds £50,000 (1981 prices) — NO / YES

Works are of a complex nature or involve a high degree of building services or other specialist works or duration exceeds 12 months — YES / NO

Use JCT Standard Form Private with Quantities (80/PW) modified by Contractors Design Portion Supplement (81/CD/DP)

Private Employer

:OR

Use JCT Standard Form Local Authorities with Quantities (80/LAW) modified by Contractors Design Portion Supplement (81/CD/DP)

Local Authority Employer

Works are sufficiently complex to require a Bill of Quantities or the Architect wishes to Nominate sub-contractors or suppliers — NO / YES

Works are of a complex nature that can best be described to the tendering Contractor by use of Bills of Quantities — NO / YES

EITHER OR

Use JCT Agreement for Minor Building Works (MW/80)

Local Authority or Private Employer

Use JCT 80 Standard Forms Local Authorities without Quantities (80/LAWO)

Local Authority Employer

:OR

Use JCT 80 Standard Forms Private without Quantities (80/PWO)

Private Employer

Use JCT Intermediate Forms of contract (IFC/84)

Local Authority or Private Employer

Use JCT 80 Standard Forms Local Authorities with Quantities (80/LAW)

Local Authority Employer

:OR

Use JCT 80 Standard Forms Private with Quantities (80/PW)

Private Employer

Use JCT Fixed Fee Form of Prime Cost Contract

Local Authority or Private Employer

:OR

Use JCT Standard Forms Local Authorities with approximate Quantities (80/LAA)

Local Authority Employer

:OR

Use JCT Standard Forms Private with approximate Quantities (80/PWA)

Private Employer

Produced by the BEC 82 New Cavendish Street, London W1M 8AD. Telephone 01-580 5588. Telex 265763

Index